Metrology of the leak detection
- Practical guide -

Developed by the EMRP IND 12 consortium
October 2015

Contributors

M. Vičar, D. Pražák - CMI, Czech Republic

M. Bergoglio, D. Mari, S. Pasqualin - INRIM, Italy

F. Boineau, D. Bentouati - LNE, France

U. Becker, K. Jousten - PTB, Germany

B. Ongun, R. Kangi - UME, Turkey

W. Grosse-Bley - Inficon Gmbh, Germany

C. Laganà - Lazzero Tecnologie srl, Italy

B. Farangis - THM, Germany

Machine for leak testing of compressors - Courtesy of Lazzero tecnologie srl - Torino-Italy

LEXITIS ÉDITIONS - 19 rue Larrey, 75005 Paris – **www.LexitisEditions.fr**

© Lexitis Éditions – Imprimé en Union Européenne – Dépôt légal : février 2016
ISBN : 978-2-36233-164-0

Introduction

Leaks have not always been a major issue in vacuum technology where they are a limiting factor to the ultimate pressure and the purity of a process gas that can be reached in a vacuum vessel, but also in any other container, be it for operational reasons (e.g. engines, air bags or pace makers), security reasons (e.g. for poisonous or radioactive materials) or environmental reasons. Limitations of refrigerant gas leakages were pointed out in regulations to meet recommendations of the Kyoto Protocol (1997). The above mentioned examples are far from being exhaustive and the leak measurement is a crucial concern in many industries.

This practical guide deals with tools and theory in the field of the gas leak detection under the angle of metrology, in a range from 1×10^{-10} Pa·m^3·s^{-1} to 1×10^{-4} Pa·m^3·s^{-1}, considering leaks towards atmosphere or vacuum.

Leaks are, in general, very low gas flows. The Chapter I introduces the gas flow metrology and contains definitions and current units found in industry and research. The primary standards for low gas flows in the European National metrology institutes, which took part in the project EMRP IND12, with the associated measurement uncertainties are presented.

To ensure leak measurements traceability, people in industry need tools more practical than primary standards: leak artefacts are the secondary standards. Depending on the artefact type (permeation or conductance), the downstream pressure where the gas flows (vacuum or atmosphere), the behaviour of the instruments differs. The Chapter II gives theory and characteristics of the different leak artefacts.

The Chapter III delves into helium leak detection, a method used world-wide to detect and quantify leaks. After a complete description of helium leak detectors and their principle, the metrological characteristics of three of these detectors assessed during the project EMRP IND12 are presented.

The Chapter IV focuses on the refrigerant leak detection. The environmental issue of greenhouse gases emissions was pointed out during the Kyoto protocol in 1997 and it was then decided to limit their emission. In Europe regulations were published afterwards, where it is in particular mandatory for owners of air conditioning equipment to have a suitable refrigerant leak detector. In this chapter, principle of the refrigerant leak detectors and the evaluation of their performance according to the standard EN 14624:2012 are presented. Evaluation of eight commercially available instruments was performed in the frame of the project EMRP IND12 which has resulted in a series of recommendations for using the instruments as properly as possible.

Finally, the reader will find in the Annexes some additional information about the vocabulary of metrology, some uncertainty assessment general considerations and a simple model to predict the gas flow in capillary leaks from few measurements with one gas species.

This work underlying this guide and also the development of the guide itself was supported by the European Metrology Research Programme (EMRP) in the designated project IND12 "Vacuum for production environments". The EMRP is jointly funded by the EMRP participating countries within EURAMET and the European Union.

CHAPTER I.   Gas flow metrology: general concepts

I.1. Definitions and units

The flow of a fluid (in our case gas) through a tube lead to a gas flow q defined as the ratio of the variation of the quantity dM to the time dt flowing throught the cross sectional area A of the tube:

$$q = dM/dt \tag{1.1}$$

The quantity dM of the gas can be measured as variation of the number of moles dN, or of an element of mass dm or of an element of volume dV or the variation of the gas energy $d(pV)$ during time dt.

The following definitions shed light on the quantities and associated units.

I.1.1. Molar flow rate q_{mol}

The molar flow rate q_{mol} strictly and completely defines the amount of the gas flowing through a given section:

$$q_{mol} = \frac{dN}{dt}, \tag{1.2}$$

where dN is the number of moles crossing the section during dt. So the unit of q_{mol} is mol/s.

I.1.2. Mass flow rate q_m

The mass flow rate q_m can be determined by measuring the mass loss dm during a time dt due to the gas flow (from a reservoir for example).

$$q_m = \frac{dm}{dt}. \tag{1.3}$$

This quantity is also preferred in regulations dealing with pollution of refrigerant gases, where accepted leakages are given in a non-SI unit, the grams per year denoted $g \cdot a^{-1}$ in the standard EN 14624:2012.

If the composition of the gas is known, then so is its molar mass M, and q_m can be linked to the molar flow rate q_{mol} by the equation:

$$q_{mol} = \frac{q_m}{M}. \tag{1.4}$$

I.1.3. Volumetric flow rate q_V

The quantity of gas enclosed in a given volume depends on both the temperature and pressure of the gas within it. For this reason, it is compulsory to specify temperature T_0 and pressure p_0 at which the gas is referred to, when stating the volumetric flow rate q_V, which is then defined by the volume of gas dV flowing during time dt, at T_0 and p_0:

$$q_V = \frac{dV}{dt} \text{ at } T_0 \text{ and } p_0. \tag{1.5}$$

A very common unit for low volumetric flow rates is the standard cubic centimetre per minute (sccm), where the word standard stands for the normal conditions of temperature and pressure that are $T_0 = 273.15$ K (0.00 °C) and $p_0 = 101\,325$ Pa.

It is also possible to link q_{mol} with q_V for a gas at the temperature T and at pressure p for an ideal gas:

$$q_{mol} = q_V \frac{1}{R} \frac{p}{T} \tag{1.6}$$

I.1.4. Throughput q_{pV}

The throughput q_{pV} is the practical quantity used in the field of vacuum. It is defined as the variation of the gas energy $d(pV)$ during the time dt, for a fixed temperature T:

$$q_{pV} = \frac{d(pV)}{dt}, \text{ at temperature } T. \tag{1.7}$$

The SI unit is the Pa·m³·s⁻¹ but the mbar·l·s⁻¹ or the Pa·l·s⁻¹ are also commonly used. Thus, throughput can be determined by pressure or volume variations measurements. For example, if the pressure in a 1 litre volume rises by 10^{-3} mbar in one second at a constant temperature T, the leak rate is 10^{-3} mbar·l·s⁻¹ at T.

One can link q_{pv} with q_{mol} by making the assumption that the considered gas is ideal. Under usual conditions of leak detection (ambient temperature and pressure lower than 1013 hPa), this leads to an approximation much lower than sought after uncertainties for leak measurements. By deriving the ideal gas law:

$$pV = NRT, \tag{1.8}$$

(p, V, T, N are respectively the pressure, the volume, the temperature, the number of mole of gas and R the molar gas constant), the following relation is established:

$$q_{mol} = \frac{q_{pV}}{RT}, \text{ at the temperature } T. \tag{1.9}$$

Note: if a constant amount of gas flowing per unit time across a given section is considered at a temperature T_1 or T_2, the relation between the corresponding throughputs q_{pV1} and q_{pV2} will be:

$$\frac{q_{pV1}}{T_1} = \frac{q_{pV2}}{T_2}. \tag{1.10}$$

I.2. Conversion table for flow rate units

The Table 1 gives the conversion factors between the different units, calculated from equations (1.3), (1.6) and (1.9). The conditions of pressure, temperature and gas specie are defined when necessary.

Table 1. Conversions factors between several gas flow rate units
Example for the reading (see framed cell): 1 mol/s = 2437 Pa·m³/s at the temperature of 20 °C;

	mol/s	g/s N_2	g/s He	g/a R-134a	Pa·m³/s $T = 20$ °C	Pa·m³/s $T = 23$ °C	mL/min $T = 20$ °C $p = 101325$ Pa	L/s $T = 20$ °C $p = 101325$ Pa
mol/s	1	28.013	4.003	3.218×10^9	2437	2462	1.345×10^6	22.41
g/s N_2	0.03570	1	-	-	87.01	87.90	4.801×10^4	0.8001
g/s He	0.2498	-	1	-	609.0	615.2	3.360×10^5	5.600
g/a R-134a	3.108×10^{10}	-	-	1	7.575×10^{-7}	7.653×10^{-7}	4.180×10^{-4}	6.966×10^{-9}
Pa·m³/s $T = 20$ °C	4.103×10^{-4}	0.01149	1.642×10^{-3}	1.320×10^6	1	1.010	5.518×10^2	9.196×10^{-3}
Pa·m³/s $T = 23$ °C	4.061×10^{-4}	0.01138	1.626×10^{-3}	1.307×10^6	0.9899	1	5.462×10^2	9.103×10^{-3}
mL/min $T = 20$ °C $p = 101325$ Pa	7.436×10^{-7}	2.083×10^{-5}	2.976×10^{-6}	2.393×10^3	1.812×10^{-3}	1.831×10^{-3}	1	1.667×10^{-5}
L/s $T = 20$ °C $p = 101325$ Pa	0.04461	1.250	0.1786	1.436×10^8	108.7	109.9	6.000×10^4	1

Notes: T and p are respectively the temperature and pressure defined for the gas,
R-134a is the refrigerant 1,1,1,2-tetrafluoroethane (CH_2FCF_3),
g/a stands for a gram per year,
1 year = 31536000 seconds,
R = 8.31447 J/(mol·K).
Molar mass N_2 = 0.028013 kg/mol   Molar mass He = 0.0040026 kg/mol   Molar mass R-134a = 0.10203 kg/mol

I.3. Techniques for leak detection

Leak detection with tracer gases can be divided into two main categories corresponding respectively to qualitative and quantitative approaches: leak location and leakage measurement. In the frame of the present document, only the latter technique will be considered, in particular helium leak detection in vacuum mode and refrigerant leak detection in sniffing mode.

I.3.1. Helium detection in the vacuum mode

The element to be tested for leak-tightness is connected to the helium leak detector and pumped down below 0.01 Pa. If helium (the tracer gas) is sprayed close to the leak-tightness failure, it gets sucked in and detected by the leak detector (Fig. 1). The output signal of detector is proportional to the pumped quantity of helium and is given directly in Pa·m³·s⁻¹ or equivalent throughput units.

The detector is a simplified mass spectrometer tuned to the helium mass (detailed in the chapter III).

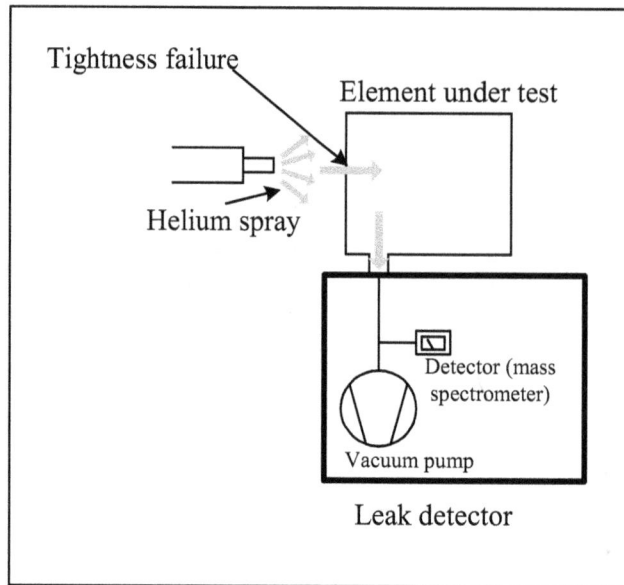

Fig. 1.1. Helium leak detection under vacuum conditions

I.3.2. Refrigerant leak detection in the sniffing mode

Refrigerant gas detection is performed at atmospheric pressure and is applied directly to the air conditioning equipment filled with the refrigerant gas under pressure. The nose of the detector, also called the sniffer is slowly moved along parts where leaks are likely to occur. The sucked gas passes through a detector sensitive to the refrigerant gas as illustrated in Fig 1.2, for a leakage existing at the connection of two pipes of an air conditioning equipment. The ultimate leak that can be detected with this technique is in the decade of 10^{-7} Pa·m^3·s^{-1}. Refrigerant gas detection is detailed in the chapter IV.

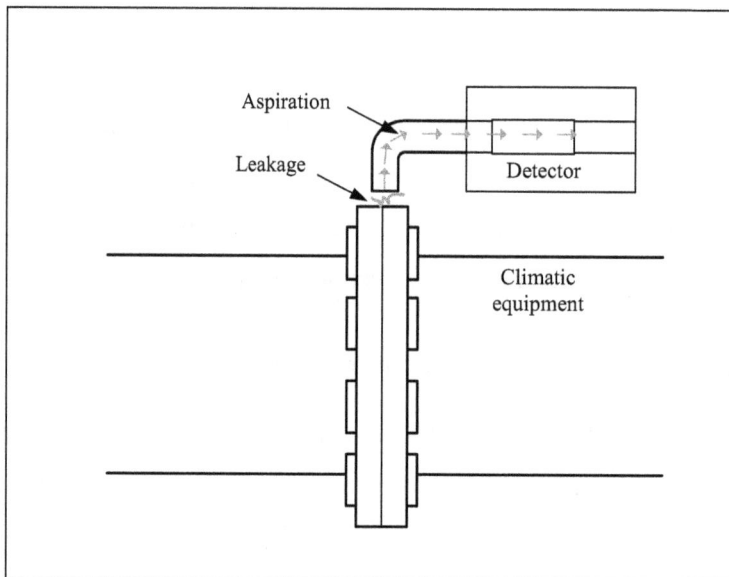

Fig. 1.2. Refrigerant leak detection in the sniffing mode

I.4. Metrological traceability of leak detection

Since both the techniques of helium leak detection in vacuum and refrigerant leak detection in sniffing mode are based on measurements, a metrological traceability chain has to be followed. Metrological traceability is defined as the "property of a measurement result whereby the result can be related to a reference through a documented unbroken chain of calibrations, each contributing to the measurement uncertainty." [1].

This traceability is illustrated in the following Fig. 1.3.

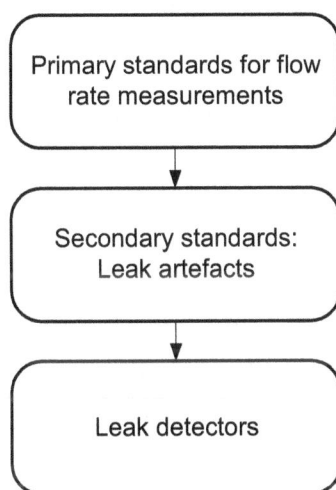

Fig. 1.3. Metrological traceability chain for leak detectors

Leak detectors are not directly calibrated with the primary standards, generally one or several transfer standards are forming the unbroken chain in between.

I.5. Primary standards

Over the years, several National metrology institutes (NMI) in the world have developed, the primary standards to measure small gas flow based on different methods. In Europe, primary standards have been developed and maintained at: CMI (Cesky Metrologicky Institut) in Czech Republic [2,3], INRIM (Istituto Nazionale di Ricerca Metrologica) in Italy [4,5], LNE (Laboratoire national de métrologie et d'essais) in France [6,7] and at PTB (Physikalisch Technische Bundesanstalt) in Germany [8,9]. The primary standards allow calibration of gaseous leaks flowing to vacuum and/or atmosphere at a high level of accuracy and are based on different methods: constant pressure flowmeter, constant volume flowmeter, photoacoustic and laminar flow element methods.

Constant pressure and variable-volume method

The method allows flows determination through the time derivative of the ideal gas equation ($p_M V_M = n R T$), that relates, through the ideal gas constant R, the number n of gas moles, contained in a measurement volume V_M (at a temperature T), to the volume V_M itself and to the pressure p_M in it. In

Fig. 1.4 a general scheme of a primary flowmeter is shown: the volumes V_R and V_M are the reference and measurement volumes respectively, while V is the inlet or outlet volume of the flow (for gas flows referred to vacuum or atmospheric pressure, respectively). In the initial configuration of the system, V_R and V_M are connected to each other, at the same pressure $p_M = p_0$; V is isolated and maintained at the pressure p (with $p \neq p_M$). At the instant t_0, in quick succession, V_M is separated from V_R by closing the valve V_a, and connected to V, by opening valve V_B. The pressure difference at the ends of the leak generates a gas flow from V_M to V (if $p_M > p$), or vice versa. The two possible types of flow that are currently measured can be summarized as follows:

- flows referred to vacuum ($p_M > p$): p_M (that is the upstream pressure of the leak) can be varied from about 70 Pa up to a maximum of about 10^5 Pa, while the downstream pressure p has a value negligible to the upstream pressure

- flows referred to atmospheric pressure ($p_M < p$): p_M (that is the downstream pressure of the leak) is about 100 kPa, while p (that is the upstream pressure) can be varied from about 1.2×10^5 Pa up to about 2×10^5 Pa.

In practice, the molar flow q_m, coming out of the volume V_M through the standard leak or coming in V_M, according to the type of flow measurement, if referred to vacuum or atmospheric pressure, respectively, is measured varying the volume V_M itself in such a way to maintain its internal pressure at a constant value p_0. Therefore, under the hypothesis that the temperature T is constant, the following relation can be directly obtained by differentiating the ideal gas law with respect to time:

$$q_m = \frac{dn}{dt} = \frac{p_0}{RT} \cdot \frac{\Delta V_M}{\Delta t} = \frac{q_{pV,p}}{RT},$$

(1.11)

where $q_{pV,p}$ is the throughput rate measured by constant pressure – variable volume flowmeter.

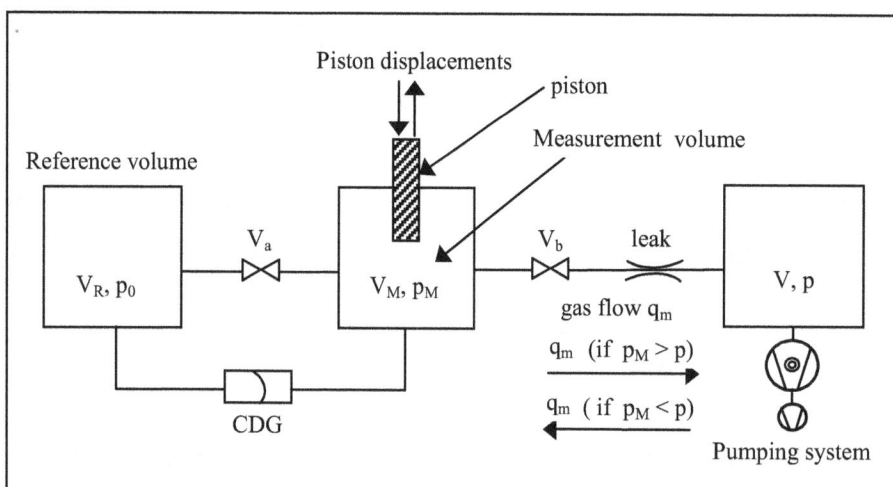

Fig. 1.4. Scheme of a flowmeter based on constant pressure and variable-volume method

The pressure difference (p_M - p_0), measured with the capacitance diaphragm gauge (CDG), is used to control the piston displacement (up or down depending on the flow direction) in order to maintain constant the pressure in the measurement volume.

Fig. 1.5. INRIM flowmeters based on constant pressure and variable-volume method

Constant volume and variable pressure method (pressure rise method)

In the case of the constant volume – variable pressure method a formula similar to (1.11) is used to measure the molar gas flow rate q_m, which is determined measuring the variation of pressure as function of time inside the constant volume V_M; under the hypothesis that the temperature T is constant, the equation (1.12) can be directly obtained time-deriving the ideal gas law:

$$q_m = \frac{V_M}{RT} \cdot \frac{\Delta p_M}{\Delta t} = \frac{q_{pV,V}}{RT} ,$$ (1.12)

in which $q_{pV,V}$ is the throughput rate measured by constant volume method and the volume V_M was previously determined by a different experiment.

As an example, a relatively easy method to measure the volume V_M consists in determining a throughput rate by constant pressure-variable volume method; then the same throughput is measured by pressure rise method and comparing the two results, it is possible to calculate the volume V_M.

Photoacoustic method

As the refrigerant gases absorb at wave lengths in the infrared range, infrared detection is particularly suitable to measure the concentration of a refrigerant gas. The method developed at LNE [7] consists in measuring the accumulation of the gas emitted by the refrigerant leak in an enclosed "accumulation volume" under atmospheric pressure. The rise of the refrigerant gas concentration is then measured by an infrared photoacoustic spectrometer.

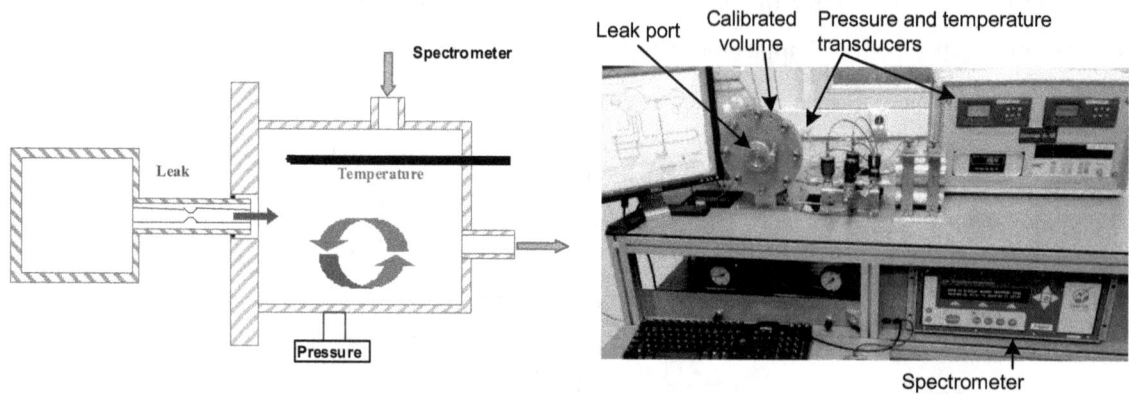

Fig. 1.5. Scheme and photograph of the system to measure gas flows supplied through the leak in calibration

The flow rate can be calculated using the equation:

$$q_m = \frac{MV}{R} \cdot \frac{\partial}{\partial t}\left(\frac{p \cdot C}{T}\right),$$ (1.13)

where M is the molar mass of the gas, V is a volume calibrated by the static expansion method, R is the ideal gas constant, p and T are the pressure and temperature inside the volume V, C is the gas concentration measured by the infrared photo-acoustic spectrometer.

Laminar flow element method

The laminar flow element measures the mass flow of the gases indirectly by measuring the temperature and the inlet and outlet pressures. The pressure difference across a pipe is directly proportional to the flow rate. Laminar flow conditions are present in a gas when the Reynolds number of the gas is below the critical figure. The viscosity of the fluid is compensated.

The laminar flow elements used are constructed as a cylindrical interstice between inner and outer cylinders to achieve the required flow rate. The laminar flow elements are employed as the secondary standards with direct traceability to the primary gravimetric flow standard (GFS). The GFS principle is based on the measurement of the weight loss of the gas in the bottle. This makes it possible to ensure traceability for different gases.

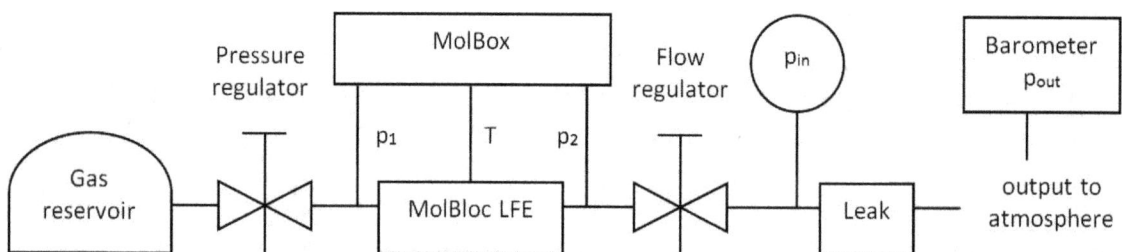

Fig.1.6. Schematic of laminar flow element method. The gas from the reservoir is passing through MolBloc laminar flow element and through the leak to the atmosphere. The leak inlet and outlet pressures are measured using two absolute pressure gauges.

In the table 1.2 the methods and related uncertainty of the participating national metrology institutes are listed.

Table 1.2. Uncertainties of NMIs facilities

NMI	Method	Gas flow range (mol·s^{-1})	Uncertainty ($k = 2$)
CMI	Constant pressure method (vacuum)	$4\times10^{-12} \div 3\times10^{-5}$	8×10^{-14} mol s^{-1} + $1.2\times10^{-2}\times q$
CMI	Laminar flow elements (atmosphere)	$7\times10^{-8} \div 2\times10^{-2}$	0.3 %
INRIM	Constant pressure method (atmosphere)	$4\times10^{-10} \div 2\times10^{-7}$	4.6% $\div$ 0.4%
INRIM	Constant pressure method (vacuum)	$4\times10^{-10} \div 2\times10^{-7}$	2.1% $\div$ 0.3%
LNE	Constant pressure method (atmosphere)	$4\times10^{-10} \div 8\times10^{-7}$	9×10^{-11} mol.s^{-1} + $0.1\times10^{-2}\times q$
LNE	Photo acoustic method (R134-a atmosphere)	$3\times10^{-10} \div 2\times10^{-8}$	2% to 5%
LNE	Constant volume method (vacuum)	$8\times10^{-11} \div 2\times10^{-6}$	1% to 4%
PTB	Constant pressure method (vacuum)	$1\times10^{-14} \div 4\times10^{-9}$	5.0% to 1.2%
PTB	Constant pressure method (atmosphere)	$< 7\times10^{-9}$ mol/s	1% to 2%
PTB	Pressure rise method (atmosphere)	$> 3\times10^{-9}$ mol/s	10%

I.6. Secondary standards

The secondary standards used to calibrate helium leak detectors (denoted HLD) or refrigerant leak detectors (denoted RLD) are the so-called leak artefacts which are devices that generate a known flow rate of the gas of interest. The theory, technology and characteristics of such artefacts are developed in the following chapter.

CHAPTER II.   Leak artefacts

A leak is an artefact that delivers a gaseous flow rate. Therefore, it has to be considered as a source of gas, which can flow either to low absolute pressure (vacuum) or to atmospheric pressure, leading to different physical properties.

II.1. Operating principles of leak

Secondary standard (or reference) leaks may be divided into two categories:

- reservoir leaks, which holds the tracer gas

- non reservoir leaks to which tracer gas is supplied during testing.

Depending on conditions of use, working principle and for practical reasons, it is possible to distinguish different types of leaks: permeation leaks, conductance leaks (with elements based on a physical restriction, a fixed conductance) and porous plugs, as described in the following paragraphs.

In Figure 2.1 a schematic cross section of a leak equipped with a gas reservoir is shown; it is mainly composed of the leak element (membrane, hole, sintered material) and one or two valves. A vacuum valve is briefly used to shut off the flow in order to zero the leak detector; a second valve is used to refill the leak reservoir with gas.

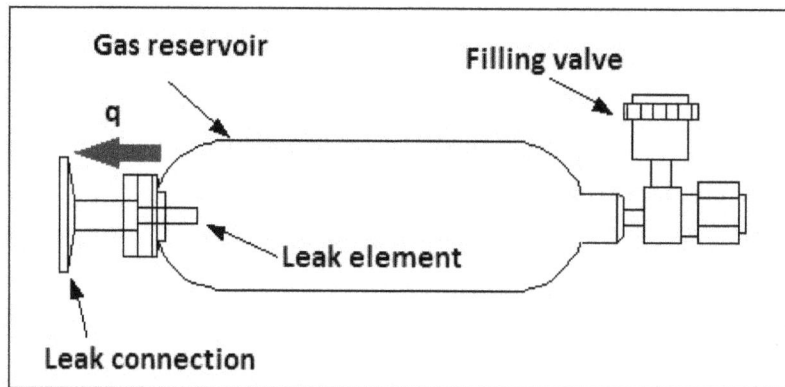

Fig. 2.1. Reservoir standard leak

II.1.1. Permeation leak

The leak element is a material permeable to the working gas (for example glass in case of helium): the tracer gas is placed in the reservoir at high concentration and permeates through the membrane wall to atmosphere or to vacuum; the gas leakage delivered by that kind of leak is based on the permeability equation of a thin membrane:

$$q = k_p A \frac{\Delta p}{L},$$ (2.1)

where q is the flow rate in Pa m^3 s^{-1}, k_p is the permeation rate constant, A is the membrane surface normal to the flow, Δp is the pressure drop along the flow path, L is the length of the flow path.

Typical gases used as tracer gas for leak testing are helium and refrigerant gases, depending on different applications. For example, a helium permeation leak is arranged in a metal reservoir with an integral glass membrane at one end. Helium diffuses through the membrane at a defined rate.

For helium leaks referred to vacuum, the generated flow range can extend from $1 \cdot 10^{-13}$ Pa·m^3·s^{-1} to $1 \cdot 10^{-4}$ Pa·m^3·s^{-1} and for R-134a leaks referred to atmospheric pressure, the flow range is typically between 1 g·a^{-1} and 50 g·a^{-1}. The gas flow delivered by the leak decreases over time (from 0.5% to 5% per year depending on the nominal gas flow and the quantities of gas in the reservoir). Experiments performed at NIST [10] have led to an empirical formula for molar flow referred to vacuum:

$$q_m = A \cdot e^{\frac{-E}{RT}}, \tag{2.2}$$

where A is a constant which depends on the solubility, the diffusivity, the surface and thickness of the permeable material (mol·s^{-1}·K^{-1}), T is the temperature (K), E is the molar activation energy of the material (J·mol^{-1}) and R is the molar gas constant (J·mol^{-1}·K^{-1}).

The temperature coefficient associated with the leak is relatively large (between 2 % and 5 % per kelvin).

II.1.2. Conductance leaks

This type of leak element consists of a restriction made on the path of the gas flow. This narrowing can be a capillary tube, an orifice or a tube with a local crimping. The conductance of the leak C depends on the geometrical characteristics of the narrowing and the average pressure across the leak element. The generated flow q_m is proportional to the conductance C and the pressure difference $(p_{in} - p_{out})$ between the inlet and the outlet of the leak.

$$q_m = C \cdot (p_{in} - p_{out}). \tag{2.3}$$

The typical range covered by this kind of leak is from $1 \cdot 10^{-8}$ Pa·m^3·s^{-1} to $1 \cdot 10^{-2}$ Pa·m^3·s^{-1}.

The artifact may be supplied with or without a reservoir filled with pure gas at a defined pressure. The leaks having the reservoir can be equipped with a pressure gauge with which the inlet pressure can be monitored; if the inlet pressure is not controlled, the pressure inside the reservoir slowly decreases and consequently also the flow rate delivered by the artifact. It is thus recommended to check the flow rate decrease, due to the gas loss in the reservoir, by means of periodic calibrations. When the inlet of the leak is fed with a constant inlet pressure, the flow rate is theoretically constant for an identical pressure difference through the leak. In addition, the user can adjust the flow rate by changing the inlet pressure.

The temperature coefficient of leak artefacts based upon a physical restriction is on the order of 0.2% - 0.6 % per kelvin.

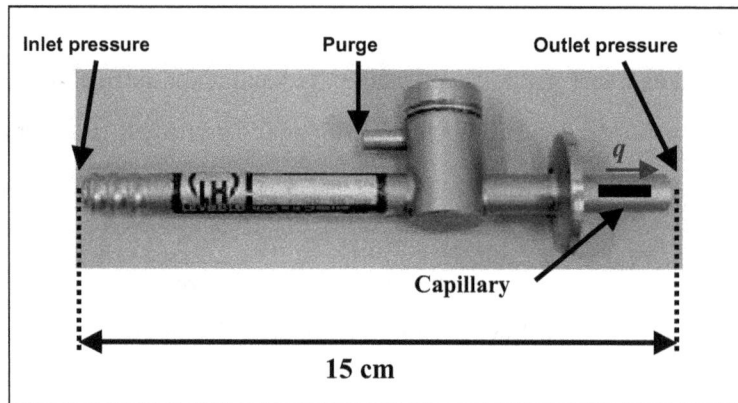

Fig. 2.2. Glass capillary Leak

II.1.3. Porous Plug leak

The leak element is a porous plug which is generally realized by means of a sintered material, to obtain a leak element consisting of several pores, with typical pore size of the order of few micrometers. A new porous body leak has been developed by the National Metrology Institute of Japan (NMIJ) and the National Institute of Advanced Industrial Science and Technology (AIST) [10]: the leak element is made of a sintered stainless steel filter with a pore size of less than 1 μm. The gas flow through the leak is molecular for an upstream pressure of less than 10^4 Pa, so for pressure below that value, the conductance can be considered constant with a relative standard uncertainty of about 3%; the dependence of the temperature is not critical with typical coefficient of about 0.2% per kelvin.

II.2. Influence parameters

The flow rate of a standard leak is influenced by various parameters, depending strongly on the type of leak, the gas species and the conditions of use. In the following paragraphs the effect of temperature and upstream and downstream pressures on the gas flow is treated as the drift over time of different types of leak.

II.2.1. Influence of temperature

Since the throughput depends on the temperature (chapter 1, equation 1.6), it is necessary to associate with the throughput value, the temperature at which it is referred. From a physical point of view, flow rates from leak elements are influenced by temperature variations in a different way depending on the type of the artefact and from gas flow regime (paragraph 2.3). If the temperature range is not too wide, the relationship between the molar flow rate and the temperature is linear and can be expressed with the following equation (2.4):

$$q_{(T)} = q_{(T_0)} \cdot \left(1 + \alpha \cdot \Delta T\right),$$
(2.4)

where $q_{(T)}$ is the flow rate measured at the temperature T, $\Delta T = (T - 273.15)$ K, $q_{(T0)}$ is the flow at the reference temperature of 273.15 K and α is the linear temperature coefficient in K^{-1}. The use of the molar flow rate is preffered to the that of the throughput as it characterises only the leak element. The behaviour of the permeation leak artefacts with temperature is well known and described in the litterature (such atype of leak is always used in vacuum conditions). Their temperature coefficients lie between 2 %·K^{-1} and 5 %·K^{-1} (§ II.1.1).

In the frame of the project EMRP IND12 [11], the temperature coefficient of several leak elements based upon a physical restriction was studied. In the Figure 2.3 the results for several gases (argon, helium and nitrogen) at several pressure differences (50, 100 and 200 kPa) are presented.

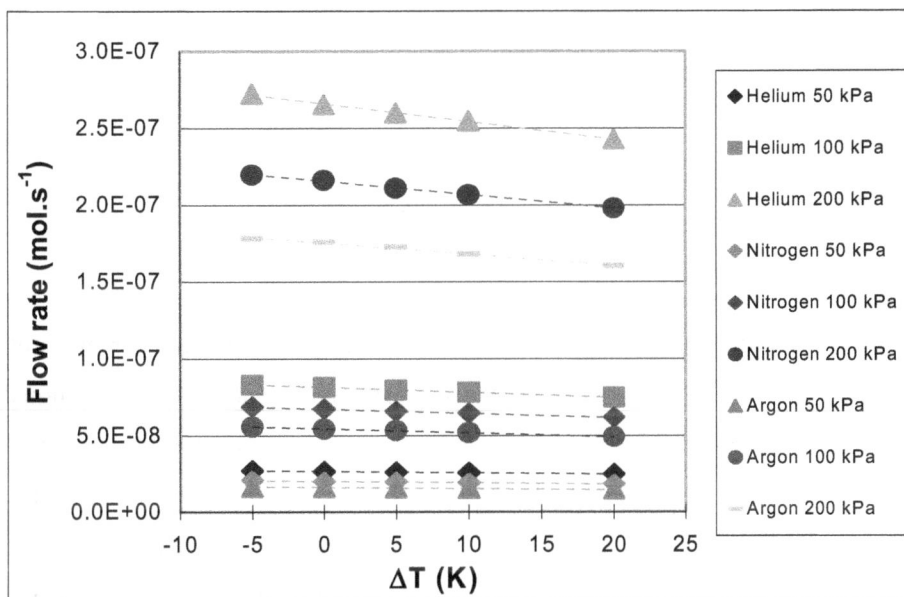

Fig. 2.3. Influence of temperature for conductance leak (argon, helium and nitrogen) at different inlet pressures

Fig. 2.4 shows the temperature coefficients α determined for a metal capillary leak for gas flows referred to vacuum and atmospheric pressure.

The temperature coefficients are around -0.4% K^{-1} for flows in vacuum and -0.6% K^{-1} for flows at atmospheric pressure for this particular leak. They were determined to be of the same order of magnitude for a sample of six metal capillary leaks. Glass capillaries show a temperature coefficient of -0.2% K^{-1}.

In the example presented in the figures 2.3 and 2.4, the temperature coefficients are typically around -0.4% K^{-1} for flows referred to vacuum and -0.6% K^{-1} for flows at atmospheric pressure. They were determined to be of the same order of magnitude for a sample of six metal capillary leaks.

Glass capillaries show a temperature coefficient of -0.2% K^{-1}.

As mentioned in the paragraph 2.1.3, in case of porous body leak, a typical coefficient of 0.2 % per K has been observed [12].

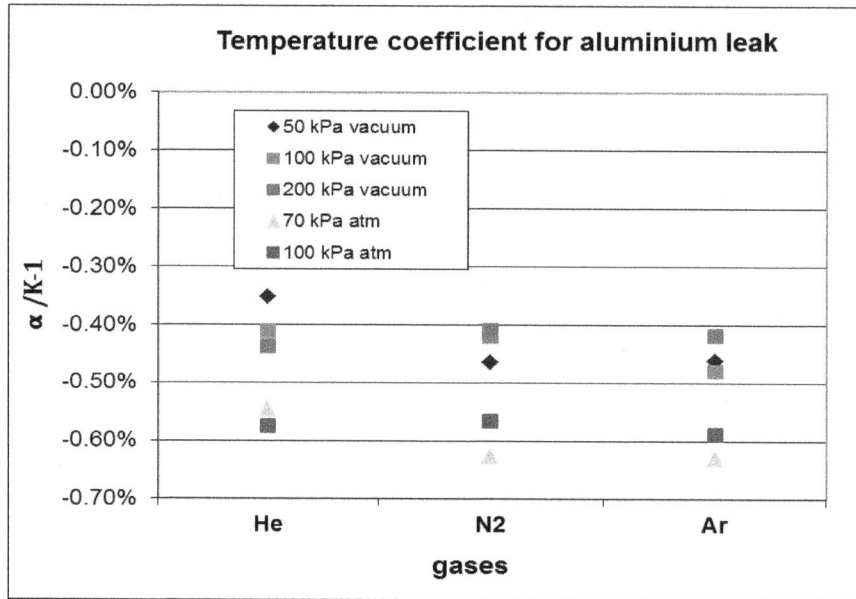

Fig. 2.4. Temperature coefficient of a metal capillary leak

II.2.2. Drift over time

The drift over time of a leak's flow rate occurs systematically when the leak element is mounted on a closed reservoir with a certain initial amount of gas. As the gas flows continuously from the reservoir, the flows rate naturally decreases. Fig. 2.5 shows the drift of the throughput at 20 °C of a permeation leak. At the time of the first calibration, the flow rate of the leak was $2.21 \cdot 10^{-10}$ Pa·m^3·s^{-1}, while seven years later it had fallen to $1.70 \cdot 10^{-10}$ Pa·m^3·s^{-1}. Since the decrease is constant, it can be evaluated by a linear model. The drift over time in this particular case is estimated to be 3.2% per year.

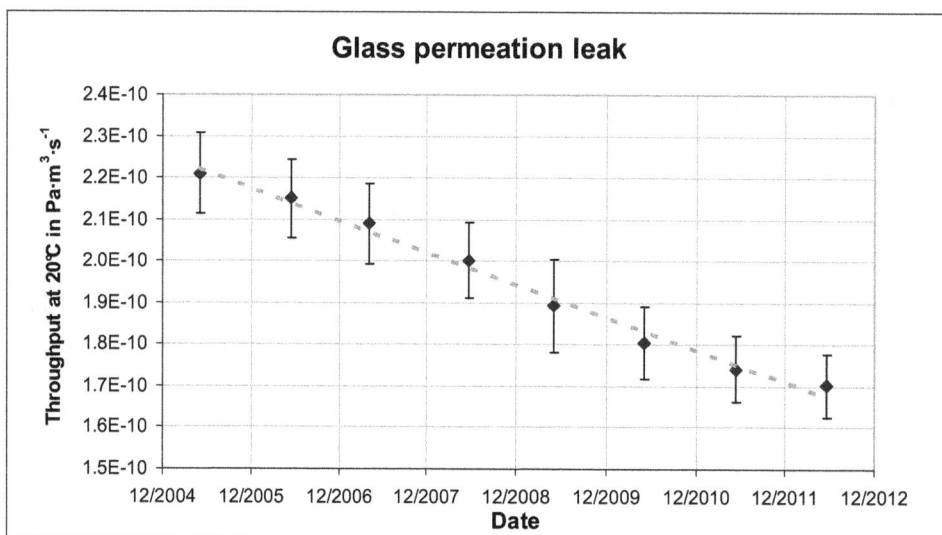

Fig. 2.5. Drift of the throughput over time for a permeation leak, vertical bars represent the calibration uncertainty ($k = 2$)

Fig. 2.6 shows the successive calibration results of a capillary leak referred to vacuum. The inlet pressure of helium is monitored to 150 kPa during the calibration. The drift over time can be considered as a random phenomenon. The experimental standard deviation of the plotted results is, in relative value, 1.0 %. The maximum drift stated between two consecutive calibrated flow rates is 1.6 % in relative value.

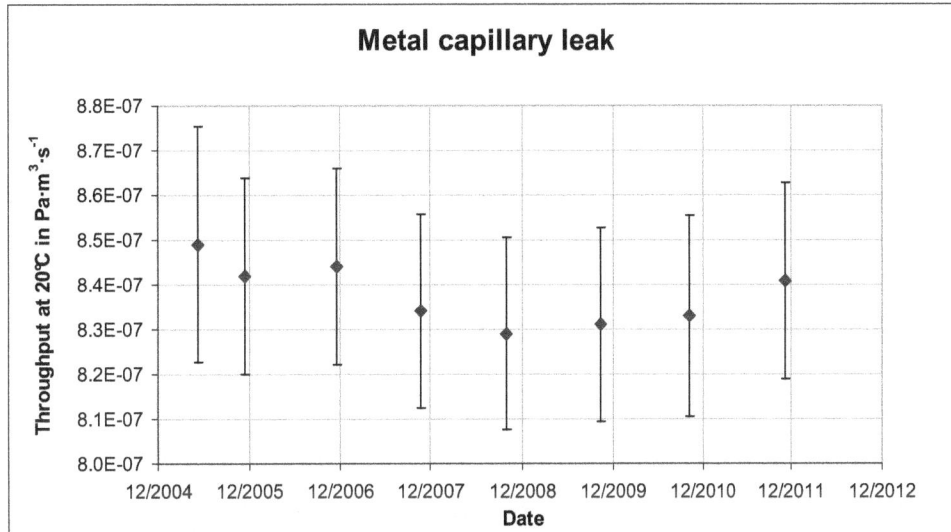

Fig. 2.6. Drift over time of a metal capillary leak referred to vacuum; the inlet pressure is 150 kPa. vertical bars represent the calibration uncertainty ($k = 2$)

Thus, the drift over time is more related to permeation leaks and can be easily estimated by periodic calibration. The other type of leak (based on physical restriction or porous body) shows a random variation over time unless partial occlusion of leak element which may result in a change of conductance of the leak. Also in this case a periodic calibration can detect such an eventuality and avoid errors during leak detector calibration.

II.2.3. Effect of atmospheric pressure changes on the gas flow delivered through a capillary leak

Leakage depends on the pressure difference across the leak. When a reference leak is used to set the sensitivity of the detector, the effect of the atmospheric changes must be considered. At high altitudes, where atmospheric pressure falls below than 100 kPa, the gas flow delivered by the calibrated leak is greater than the corresponding flow delivered in an environment at 100 kPa.

In the viscous flow regime, the flow is proportional to the difference of the squares of the absolute pressure:

$$q = \frac{\pi r^2}{8\mu l}\left(p_1^2 - p_2^2\right),$$ (2.5)

where q is the leakage measured, r is the radius of the capillary leak, l its length, μ is the viscosity, p_1 is the upstream pressure and p_2 is the downstream pressure.

When the downstream pressure p_2 changes to p_3, the leakage is given by:

$$q_b = \frac{\pi r^2}{8\mu l}\left(p_1^2 - p_3^2\right)$$ (2.6)

and:

$$\frac{q}{q_b} = \frac{\left(p_1^2 - p_2^2\right)}{\left(p_1^2 - p_3^2\right)},$$ (2.7)

which leads to:

$$q_b = q\frac{\left(p_1^2 - p_2^2\right)}{\left(p_1^2 - p_3^2\right)}.$$ (2.8)

In the molecular flow regime, the flow of gas through a leak is proportional to the difference between the pressures acting across the leak:

$$q = 3.342\frac{r^3}{l}\sqrt{\left(\frac{RT}{M}\right)}(p_1 - p_2),$$ (2.9)

where q is the leakage rate measured at p_1 and p_2 pressure, r is the radius of the capillary leak, l its length, R is the molar gas constant, T is absolute temperature of the leak, M is the molecular weight of gas, p_1 is the upstream pressure and p_2 is the downstream pressure.

When the downstream pressure p_2 changes to p_3 the leakage is given by:

$$q_b = q\frac{\left(p_1 - p_3\right)}{p_1 - p_2}$$ (2.10)

In the frame of the project EMRP IND12, an experimental study was carried out to demonstrate the effect of the atmospheric pressure changes; a capillary leak was connected to a primary flowmeter to measure directly the flow delivered at three different levels of the inlet pressure. The first measurements sets were recorded at atmospheric pressure of the laboratory, typically 99 kPa (outlet pressure).

In table 2.1 the flow, its expanded uncertainty, the inlet and outlet pressure values are summarized.

Table 2.1. Inlet and outlet absolute pressures values, p_{in} and p_{out}, used to generate the gas flow from the leak

p_{in} (kPa)	p_{out} (kPa)	q (Pa·m^3·s^{-1})	$U(q)/q$ %
106064	99064	1.64×10^{-5}	1.5
119120	99120	5.00×10^{-5}	1.5
149336	99336	1.38×10^{-4}	0.7

Afterwards, two pressure regulators were connected to the flowmeter: the first one allows one to maintain constant the capillary inlet pressure during the measurements. The second pressure regulator was used to change and maintain constant the outlet pressure of the leak, that, for the specified set-up shown in Fig. 2.7, corresponds to maintain a constant pressure in the reference volume of the flowmeter.

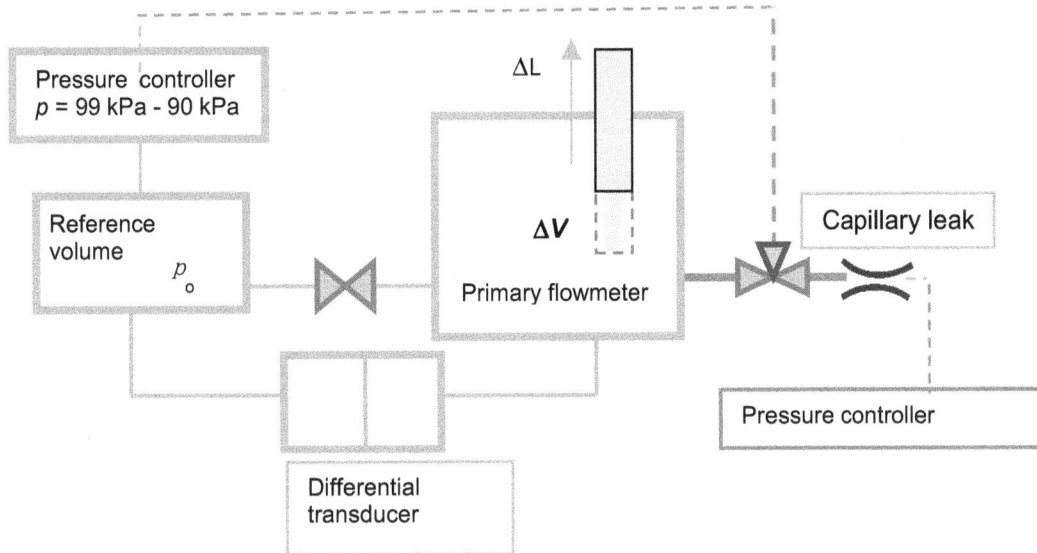

Fig. 2.7. Scheme of the primary flowmeter and leak under calibration

The outlet pressure was decreased from atmospheric value, 99 kPa, to 90 kPa. The results are summarized in figure 2.8.

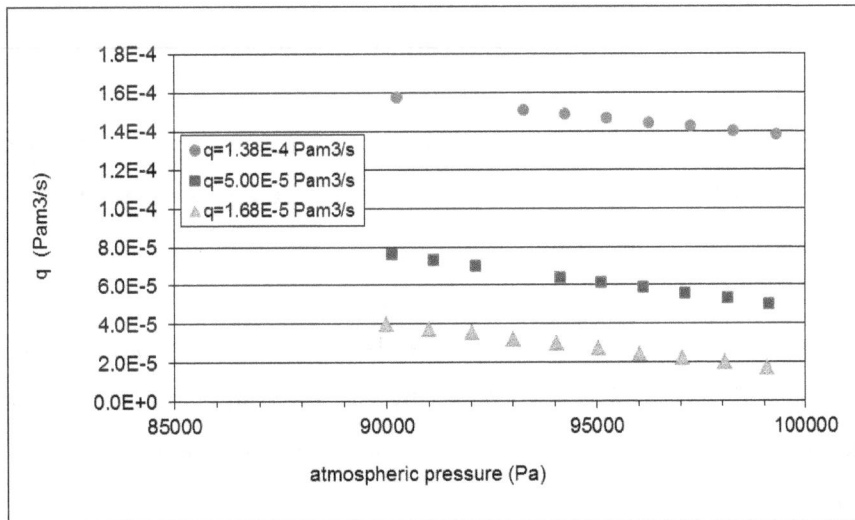

Fig. 2.8. Influence of atmospheric pressure on the gas flow delivered by a capillary leak

For gas flow 1.68×10^{-5} Pam^3/s at 99 kPa, a reduction of 1% of the atmospheric pressure corresponds to an increase of the gas flow equal to 15%.

In case of gas flow equal to 1.38×10^{-4} Pam^3/s at 99 kPa, a reduction of 1% of atmospheric pressure corresponds to an increase of the gas flow equal to 1.5%.

In figure 2.9 the trend of the relative variation of the pressure and the flow rate are reported.

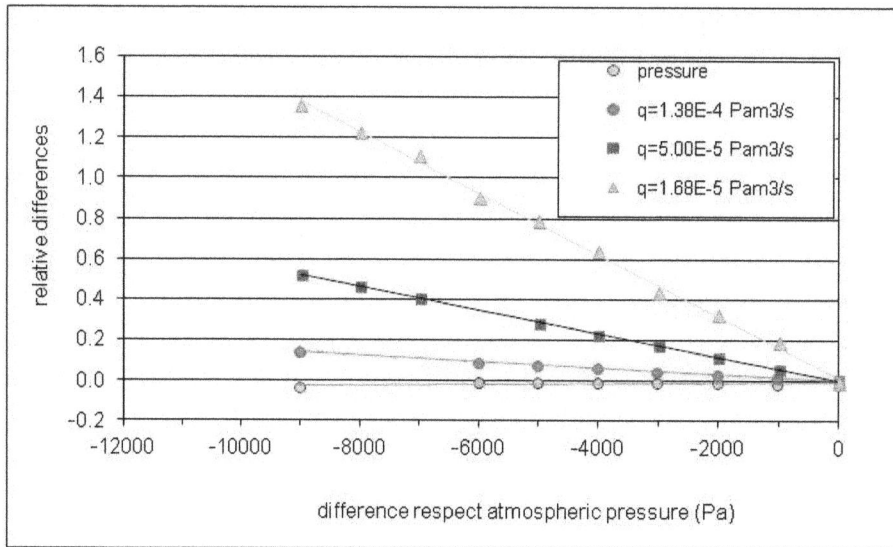

Fig. 2.9. Trend of the variation of atmospheric pressure and corresponding variation of the gas flow

From the results it can be concluded that when the atmospheric pressure decrease of 9% the flow rates increase respectively of 14%, 52 % and 136% their corresponding values at 99 kPa.

In the case of the leak having a flow rate at 99 kPa equal to 1.38×10^{-4} Pa·m^3/s, the regime is viscous and the flow increases to the values as the ratio of the square of the pressure differences. For the smaller leak $q = 1.68 \times 10^{-5}$ Pa·m^3/s, where the flow is near the molecular regime, the flow increases as the ratio of pressure differences.

In conclusion, a reference leak without its own gas supply allows one to overcome the atmospheric pressure variation because a constant differential pressure can be supplied.

II.2.4. Flow regime

The flow regimes through the leak artefact may be divided in three basic types, viscous flow, molecular flow and transitional flow, depending by the mean free path of gas molecules and geometrical characteristics of the leak element (for example the diameter and the length of a cylindrical pipe).

The flow regime may be identified by the Knudsen number *Kn,* which expresses the relation between the mean free path and the characteristic geometrical dimension of the leak element:

$$Kn = \lambda / r \qquad (2.11)$$

where λ is the molecular mean free path and r is the geometrical dimension (the radius of the tube or the height of the channel).

The Knudsen number [13] can be written as:

$$Kn = \frac{1}{r} \frac{\sqrt{\pi}}{2 p_{av}} \left(\frac{2 k_B T}{m} \right)^{1/2} \qquad (2.12)$$

22

where T is the temperature of the gas, k_B is Boltzmann constant $k_B= 1.380658 \times 10^{-23}$ J/K, m is the molecular mass of the gas, μ is the shear viscosity which can be found in [14], (annex A) and p_{av} is the average pressure across the leak.

Generally, when $Kn \gg 1$, the free molecular regime occurs. In this case, at the molecular level, the mean free path length of gas molecules is greater than the largest cross sectional dimension of the leak. The molecular regime is generally linked to a flow referred to vacuum. For a value of $Kn \ll 1$, the molecular collisions against the walls are negligible, the mean free path length of gas is smaller than the cross section of the leak and the hydrodynamic flow occurs; it is typically related to a gas flow from the leak released at atmospheric pressure. The transition regime occurs for Kn number having intermediate values where the gas cannot be considered as a continuous medium and not even in molecular regime. Summarizing, the considered flows can be described through different models: from the molecular model up to the hydrodynamic one, through the transition model.

Hydrodynamic flow coincides with the viscous flow described by Hagen-Poiseuille formula for compressible flows, when the Knudsen number tends to zero: the viscous stresses are balanced by the pressure gradient, thus neglecting inertial terms. In order to evaluate the effect of inertial terms in hydrodynamic regime the compressible Navier-Stokes equations have to be considered.

For each regime, the conductance shows a different dependence from the involved physical quantities: pressure across the physical restriction (for example a capillary) of the leak, temperature and chemical properties of the gas, geometrical dimensions of the leak itself. The well know expressions in the lower (molecular) and upper (viscous) limits are:

$$C_m = \frac{2\pi r^3}{3l} \cdot \sqrt{\frac{8RT}{\pi M}} \tag{2.13}$$

and

$$C_v = \frac{\pi r^4}{8\eta l} \cdot p_{av} \text{ (Poiseuille equation)}, \tag{2.14}$$

where C_m/C_v indicates the conductance in the molecular/viscous regime. The others symbols have the following meanings: r, l radius and length of the capillary (m); η, T, M: viscosity (Pa·s), absolute temperature (K) and molar mass (kg·mol^{-1}) of the gas; p_a: mean pressure across the capillary (Pa); R: molar gas constant (J·mol^{-1}·K^{-1}).

The description of transition regime is more complex: as part of a European research project EMRP IND12, investigations have been carried out in order to study, in particular, the transition regime of leak artefact when the applied pressure changes and different gas species are considered. Figure 2.10 shows the results for a metal capillary leak used with several gases in vacuum and at atmosphere. In order to have a significant view, the conductance C was calculated from equation (2.2), $C = q/(p_{in}-p_{out})$, and plotted as a function of the inverse of mean free path of molecules; the separation lines of the different flow regimes are shown. In molecular regime, the conductance is constant and at the same temperature, depends only on molecular mass of molecules. In viscous (or laminar) regime, the conductance depends on dynamic viscosity and on pressure in the leak element. For *Kn* between 0.01 and 0.5, the conductance depends both on molecular mass, pressure and viscosity.

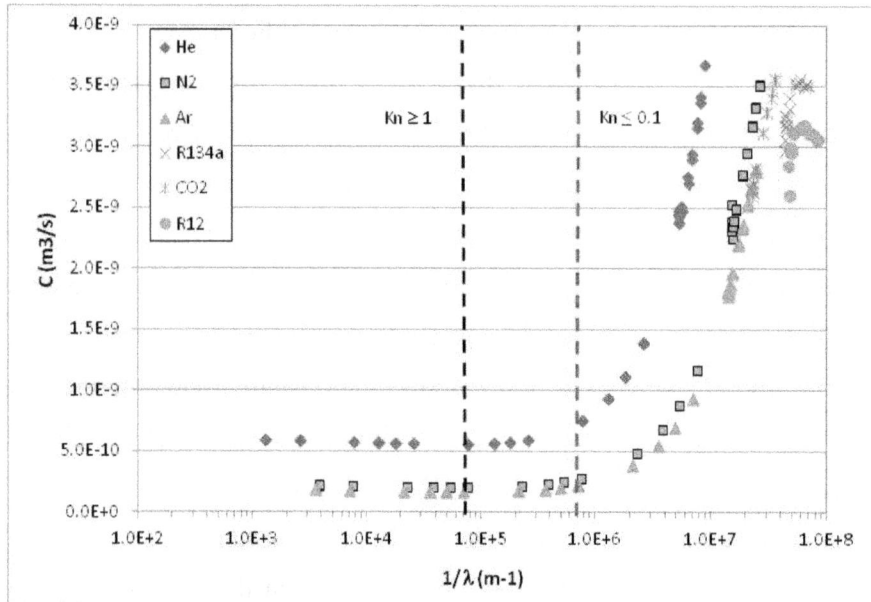

Fig. 2.10. Conductance of a capillary leak for different gases and pressure

The experimental data have been compared with the predictions of the bridging formula proposed by Gallis and Torczynski [15] with the aim of exploring the possibility of using a relatively simple way to found the geometric parameters of the leak under test on a reduced set of experimental data in order to predict gas flow rates in conditions where experimental data are not available. The formula has a simple structure and allows the computation of mass flow rates with a very small computational effort. The formula tries to bridge the two limit cases of free molecular and viscous regimes where a simple analytical treatment is available.

The complete discussion is given in Annex A.

The experimental data have been compared with the values evaluated by the equations proposed by Gallis and Torczynski [15] that can cover the whole range of rarefaction conditions.

Figure 2.11 summarizes both the experimental data and the calculated values of the normalized conductance ($C_n = C\sqrt{M}$) versus the inverse of the mean free path; in Figure 2.12 the relative differences between those data are plotted.

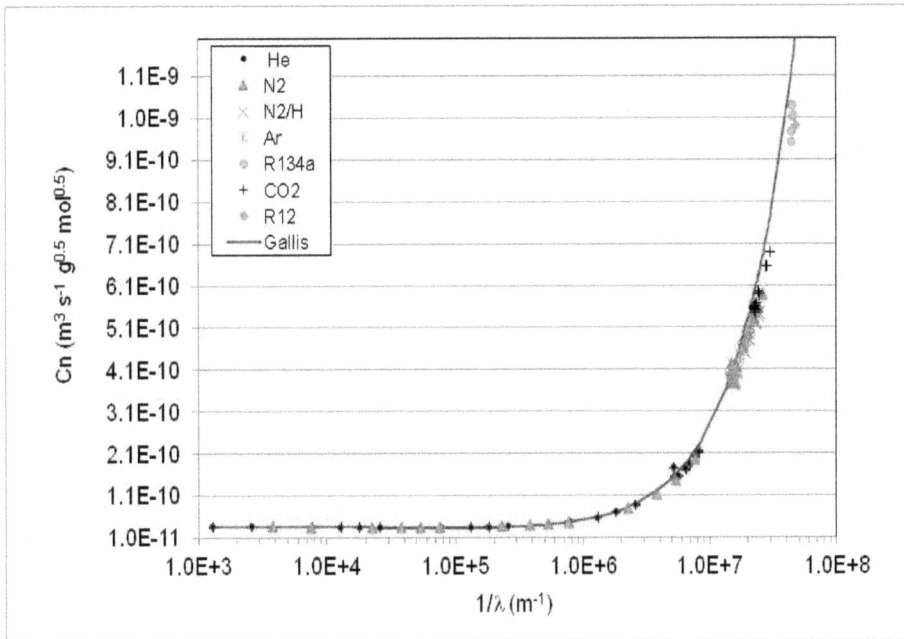

Fig. 2.11. Experimental data and calculated values of the normalized conductance versus the inverse of the mean free path.

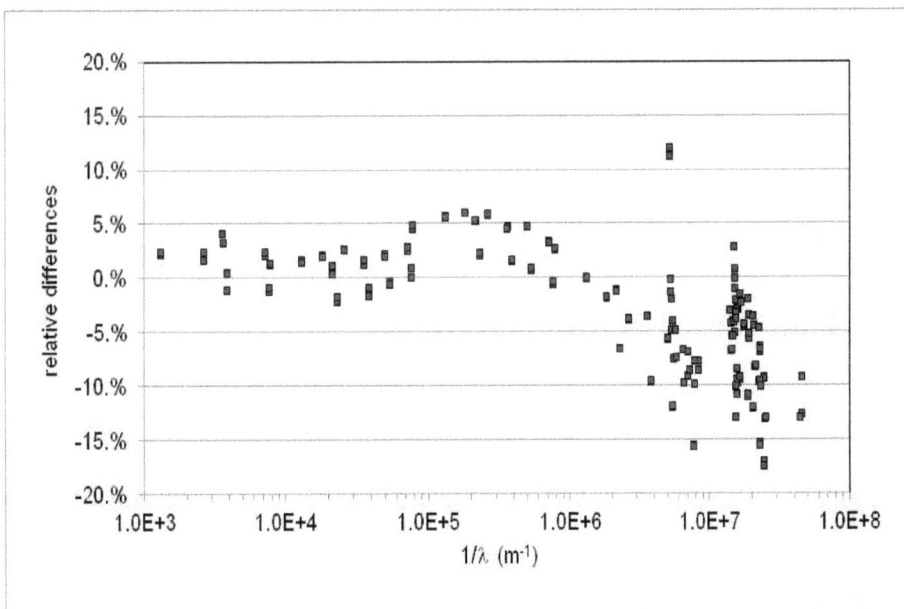

Fig. 2.12. Relative differences between those data.

The experimental data and the calculated values (coming from the bridging formula) show a maximum relative difference of 20 % when the inlet pressure of the capillary is higher than 0.1 MPa or a polyatomic gas as R-134a or R-12 is expanded at atmospheric pressure.

CHAPTER III.   Helium leak detection

III.1. Introduction

The helium mass spectrometer leak detector (MSLD) is considered the most versatile of the industrial leak testing method [16].

Leak detector history starts in the 1943, in the middle of World War II when talents in the world were engaged in developing a new type of bomb based on the fission of the uranium atom under the "Manhattan Project". Alfred Otto Carl Nier made an important contribution by developing a mass spectrograph used for monitoring uranium separations in the plant and a supersensitive leak detector.

Subsequently, Nier designed a revolutionary new mass spectrometer, based on a 60° sector field magnet, in which the ion source and detector were removed from the influence of the electromagnet. This simple design reduced the power consumption of the electromagnet (and fabrication costs) and showed better characteristics in operation to the earlier mass spectrometers without any decline in resolution.

Another feature of the Nier spectrometer was the use of electronic measurement of the ion beams, in contrast to the photographic measurement techniques used at the time.

He also helped develop helium-leak detectors having high sensitivity, because of its industrial use, the chosen material, originally glass, turned out to be unbearably fragile and after many complaints by the users, a new metallic version was developed.

III.2. Why helium is used in leak detection?

Helium-4 is an inert, non-toxic, non-condensable unreactive, colorless, and odourless monoatomic gas that is plentiful and relatively inexpensive. It is a small molecule and light, and therefore easily slips through very small leaks. Fortunately, there is only a low concentration of helium naturally present in the atmosphere (about 5 parts per million), so normally occurring background levels are manageable.

Helium was discovered independently in 1895 by Sir William Ramsay in London and by N. A. Langley and P. T. Cleve at 1895 in Uppsala, Sweden.

While there is some helium in the atmosphere, currently its isolation from that source by liquefaction and separation of air is not normally economic. This is because it is easier, and cheaper, to isolate the gas from certain natural gases. Concentrations of helium in natural gas in the USA are as high as 7% and other good sources include natural gas from some sources in Poland. For many years the USA produced over 90% of commercially usable helium in the world. By 2000, Algeria became the second largest producer of helium. In recent years, both helium consumption and the costs of helium production have increased; in the period from 2002 to 2010 helium prices doubled. Due to the

increase of helium use and the limited availability of this element on Earth, there is a risk that the terrestrial helium reserves will expire by 2040.

The fluid properties of helium: density, molar volume, internal energy, enthalpy, entropy, C_v, C_p sound speed, Joule-Thomson coefficient, viscosity, thermal conductivity, speed of sound are summarized in Annex table 1, for a temperature of 20 °C and in the pressure range from 10 kPa to 1 MPa. Properties in other conditions of temperature or pressure and for other gas species can be found on the NIST website [14].

III.3. Leak detector operation principle

Helium mass spectrometer leak detectors are available in many different configurations [17, 18, 19]. All of them consist of three parts:

- spectrometer tube,

- vacuum system,

- electronics for operating the system.

These components can differ from one manufacturer to another but they operate in a similar manner.

The central part of the helium leak detector is the cell in which the residual gas is ionized and the resulting ions accelerated and filtered in a mass spectrometer.

The most common way to have ions is to bombard gas of the sample with electrons of about 70 eV. The electrons are generated by heating a metal wire (filament), commonly made of tungsten or rhenium. A voltage of about 70 V accelerates the electrons towards the anode.

During the bombardment, one or more electrons can be removed from the neutral molecule thus producing positively charged molecular radical ions.

Only about one in 10^3 of the molecules present in the source are ionized. The ionization probability differs among substances, but it is found that the cross-section for most molecules is a maximum for electron energies from approximately 50 to 100 eV. Most existing compilations of electron impact spectra are based on spectra recorded with approximately 70 eV electrons, since sensitivity is here close to a maximum and fragmentation is unaffected by small changes in electron energy around this value. During this ionization, the radical-ions gain on average an excess energy enough to break one or more bonds and hence producing fragment-ions.

Most of the current detectors use, as in the original design, a magnetic sector to separate the helium ions from the other gases. Permanent magnets are generally used to generate the magnetic field. The adjustment needed for the selection of the helium peak is made by varying the ion energy.

In a mass spectrometer with fixed geometry (commonly used for leak detector) the ion path radius at which ions traveling along the circular path will strike the collector plate is a fixed radius r_0. In this case the equation for the ratio m/q of ion mass to charge is:

$$\frac{m}{q} = \frac{(r_0 B^2)}{2V}. \tag{3.1}$$

Equation 3.1 indicates that the particular mass of a singly ionized particle striking the collector plate, depends on the intensity of the magnetic field B and the accelerating voltage V.

For a mass spectrometer with a fixed magnetic field of intensity B, the ion mass-to-charge ratio is given by:

$$\frac{M}{e} = \frac{K_{ms}}{V}, \tag{3.2}$$

where M is the mass of positive ion given in atomic mass units (u). The charge on the positive ion is given in units equal in magnitude to the charge on one electron 1.6×10^{-19} C. The term K_{ms} is a characteristic constant of the particular mass spectrometer with fixed geometry and magnetic field intensity that is selected for use. In this way the ion mass that strikes the target depends only on the accelerating voltage V.

For example, if an instrument whose constant $K_{ms} = 1200$ and a signal peak had appeared at a voltage $V = 300$ V from equation 3.2 the ratio of the ion mass M in atomic mass units (u) to ion charge in units of electron charge e is $M/e = 4$ and helium tracer gas would have been detected.

After their separation in the magnetic field of the mass spectrometer, ions with a specific mass-to-charge ratio can be selected to strike an ion collector located beyond the slit. Each ionized ion, travelling at high speed, strikes the collector plate (Faraday cup) generating a net positive charge. Ions cause secondary electrons to be ejected. This production of electrons constitutes a temporary flow of electric current until the electrons have been recaptured. This collector current is then amplified by the electrometer stage; it is often placed within the high vacuum enclosure to ensure signal stability, minimize the electrical time constant and reduce the stray noise pickup. This amplified signal current is then typically displayed on a readout display of the mass spectrometer leak testing instrument. The Faraday cup detector is simple and robust and is used in situations in which high sensitivity is not required.

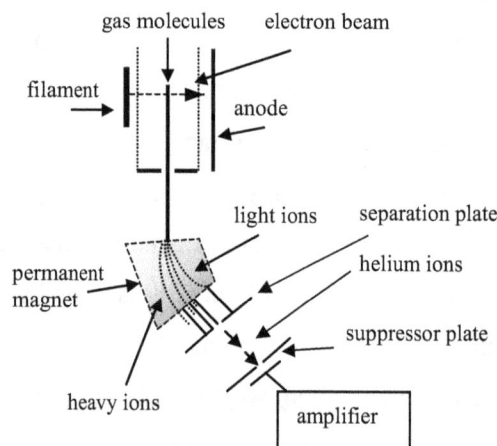

Fig. 3.1. Leak detector scheme

III.4. System configuration

III.4.1. Vacuum system and fittings

The vacuum system of the MSLD usually consist of a mechanical roughing pump, a turbomolecular pump, valves and gauges. The effect of an efficient vacuum system allows:

- to reduce the spurious background helium signal due to helium contamination of the atmosphere surrounding the test object or to ion scattering due to the pressure to high in the tube,

- to increase the minimum detectable leakage rate and to reduce the response time.

Plastic or rubber tubes, gasket greases, O-ring, have to be avoided because when exposed to high concentration tend to soak up helium becoming later source of helium outgassing.

III.4.2. Direct flow configuration

The spectrometer is connected to the high vacuum pump, the test piece is evacuated by a high vacuum pumping system down to a pressure sufficiently low [5]. The helium flow q_{He} is converted into a helium partial pressure p_{He} by:

$$p_{He} = {q_{He}}\big/{S_{He}},\qquad(3.3)$$

in which S_{He} is determined by the pumping speed of the high-vacuum pump.

In practice, the specimen in a leak detection system is contained in a volume that is evacuated by an auxiliary pump for pre-evacuating the specimen. Advantages of direct flow leak detectors include high pumping speed for helium at the inlet port, shorter response times even for larger specimen volumes and very low detection limits down to the 10^{-12} Pa·m^3/s range.

III.4.3. Counter flow configuration

In counter flow leak detectors, the total gas flow q from the specimen is carried away only by the primary pump from where it flows backwards, in counter flow, via the high vacuum pump and into the mass spectrometer. The helium flow q is transformed into a helium partial pressure p_{He} through the pumping speed of the primary pump reduced by a factor equal to the compression factor K_{He} in the high vacuum pump [20].

$$p_{He} = {q_{He}}\big/{(K_{He}S_{He})}.\qquad(3.4)$$

The main advantage of counter flow leak detectors is that they obtain measuring readiness very quickly; when the measurement starts, the pressure in the specimen is far from the desorption range. The transition from pre-evacuation to measurement is independent of the water vapor content and thus simple to be controlled according to the pressure.

Counter flow systems tend to have lower helium sensitivity compared to conventional systems, they allow a higher pressure at the inlet and are also less sensitive to contamination.

III.4.4. Sensitivity of helium leak detector

For a helium leak detector the sensitivity is general defined as the minimum partial pressure of helium that would produce the minimum detectable leakage indication. This minimum detectable leakage signal is defined as three times the magnitude of the random noise signal associated with the leak test.

The sensitivity is influenced by:

- total gas pressure and helium concentration in the detector tube. A high pressure in the leak detector (LD) gives a spurious signal due to scattering of the separated ions back into the ion collector. Even if no tracer gas is present, the amplifier will show a signal due to scattered ions.

- time duration of the test

Leak detector sensitivity depends on individual instrument and tends to become much more nearly constant below a pressure of 0.1 Pa.

III.4.5. Time response of helium leak detector

An additional factor to evaluate the sensitivity of helium leak detector is the response time. The response time of a leak detector is the time until the indicated leak rate rises to a certain fraction of full scale, typically 63% of its final value. The time response depends:

- the volume of the vacuum system
- the percentage of helium tracer gas constantly surrounding the test boundary
- the pressure in the vacuum system
- the pressure in the sensing element
- the length and size of connection
- the effective pumping speed for helium.

III.5. Study of leak detector performance

Different commercial helium leak detectors have been used to determine the main parameters which characterize the detectors and to give additional information on their performance.

In the present paragraph the experimental results obtained by National Metrology Institutes are summarized.

Before starting the measurements, the MSLDs were "warmed up" following the instrument manual: typically the detectors were switched on the day before.

The temperature of the laboratory was maintained at (20 ± 0.5) °C during the measurements.

Reference leaks, calibrated against the primary flowmeters, were used for the detectors characterization and they were connected to the MSLD port.

III.5.1. Noise, drift of the signal background and sensitivity

The noise and the sensitivity were determined according to the following procedure:

- warm up of the system,
- recording of the background signal by using a simple data acquisition (one reading per second or less)
- switching of the valve of the reference leak to inlet the helium gas flow in the detector,
- recording of the output signal till it has reached a stable value.

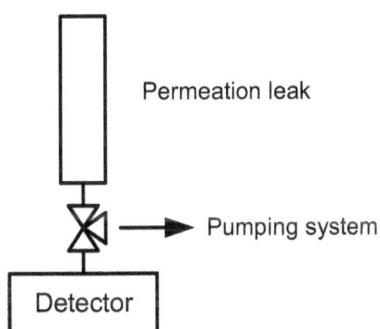

Permeation leak

Pumping system

Detector

Fig. 3.2. Experimental setup to measure noise, drift and signal background

The signal was recorded on three different days without switching off the detector.

In table 3.1 the mean values of the signal background are reported. It is represented by spurious signals given by the leak detector without the search gas and the noise, calculated as the standard deviation of the data recorded in a period of about 200 s.

The noise was, practically, for all the measurements less than 1% of the signal.

Table 3.1. Signal and background noise of MSLD measured in three different days

	1st measurement	2nd measurement	3th measurement
Signal mean value (Pa m³/s)	2.81E-11	2.87E-11	2.55E-11
Noise (Pa m³/s)	2.02E-13	1.57E-13	1.01E-13

In table 3.2 the results of the signal drift are summarized.

Table 3.2. Drift of the signal and background of MSLD measured in three different days

	drift of the signal background		
	1st measurement	2nd measurement	3th measurement
in 600 s	-17%	-3%	-3%

For the detectors considered, the signal drift of the background may be estimated, in the worst case, equal about to 2% per minute.

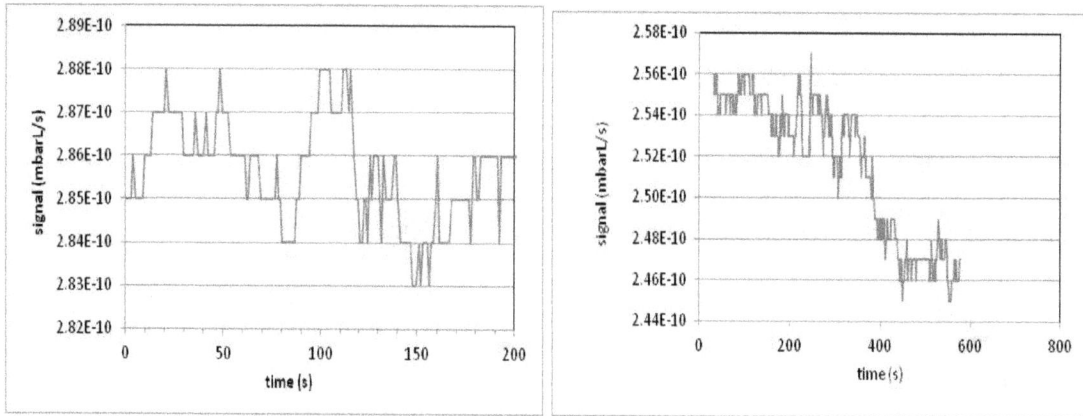

Figure 3.3. Example of noise and drift of the MSLD signal background

III.5.2. Time response

A detector was used with four different reference leaks to measure the time response. The reference leaks were connected directly to the port of the detector; the results are summarized in table 3.3 and an example is shown in figure 3.4.

Table 3.3. Reference leaks values, signal at 63% and corresponding time response.

	Leak value (Pa m³/s)	Signal at 63% of the value (Pa m³/s)	Time response (s)	Signal at 95% of the value (Pa m³/s)
Leak 1-Detector 2	8.01E-06	5.04E-06	42	7.61E-06
Leak 2-Detector 2	1.69E-06	1.06E-06	41	1.60E-06
Leak 3-Detector 2	2.54E-07	1.60E-07	25	2.41E-07
Leak 4- Detector 2	6.87E-08	4.33E-08	4	6.53E-08

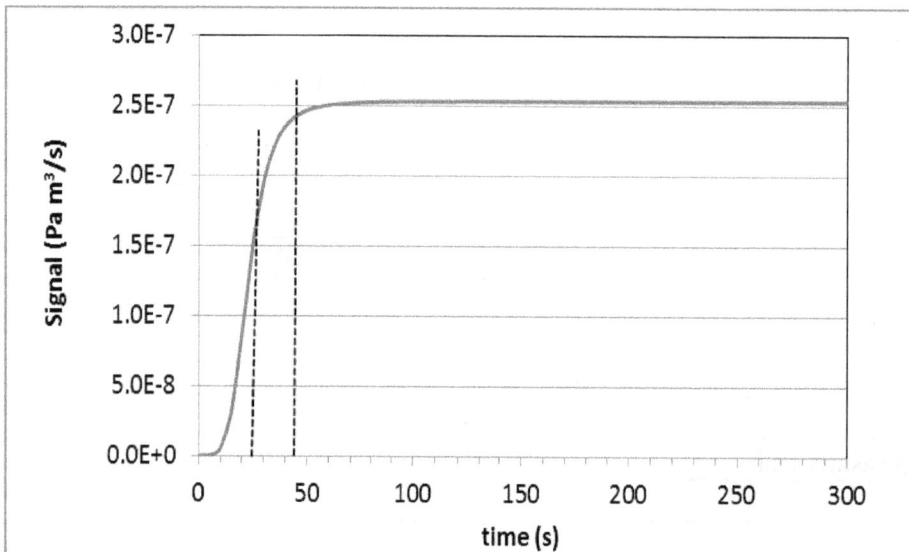

Figure 3.4. Example of MSLD signal: first dot line indicates 63% of the signal; the second dot line indicates 95% of the total flow released by the leak

III.5.3. Non-linearity, long term stability, behavior after a self-calibration

Helium leak detectors from three different manufacturers and used in the past in different environments (the first is moved and used in different places, the second was used in industrial laboratory and the last is a new detector) were calibrated by means of reference leaks on several consecutive days.

Nonlinearity, long term stability, behavior after a self-calibration are the parameters described in the following paragraph.

The experimental measurements were performed on different days following the procedure shown in table 3.4 where $t0$ is the time of the first calibration.

Table 3.4. Time plan of measurements

$t1 =$	$t0$ + 1 day	measurements from the lowest to the highest leak rate
$t2 =$	$t0$ + 3 days	measurements from the lowest to the highest leak rate
$t3 =$	$t0$ + 1 week	measurements from the lowest to the highest leak rate and then from the highest to the lowest flow rate
$t4 =$	$t0$ + 2 weeks	measurements from the lowest to the highest leak rate
$t5 =$	$t4$ + 1 day after a self-calibration	measurements from the lowest to the highest leak rate

Non-linearity

The non-linearity of the signal is defined as the deviation from the linear model. In figure 3.5 are summarized the results obtained from three different detectors.

Detector 1 shows a strong non-linearity at time $t0$ and $t1$ higher than 15%, better results are pointed out in the later measurements where the deviation from linearity is in the range -6% ÷ +6%.

Detector 2 shows a non linearity in the range from -4% to +4%.

Detector 3 shows the best characteristics of linearity, in fact the deviation from the linear model is in the range -4% ÷ 0%.

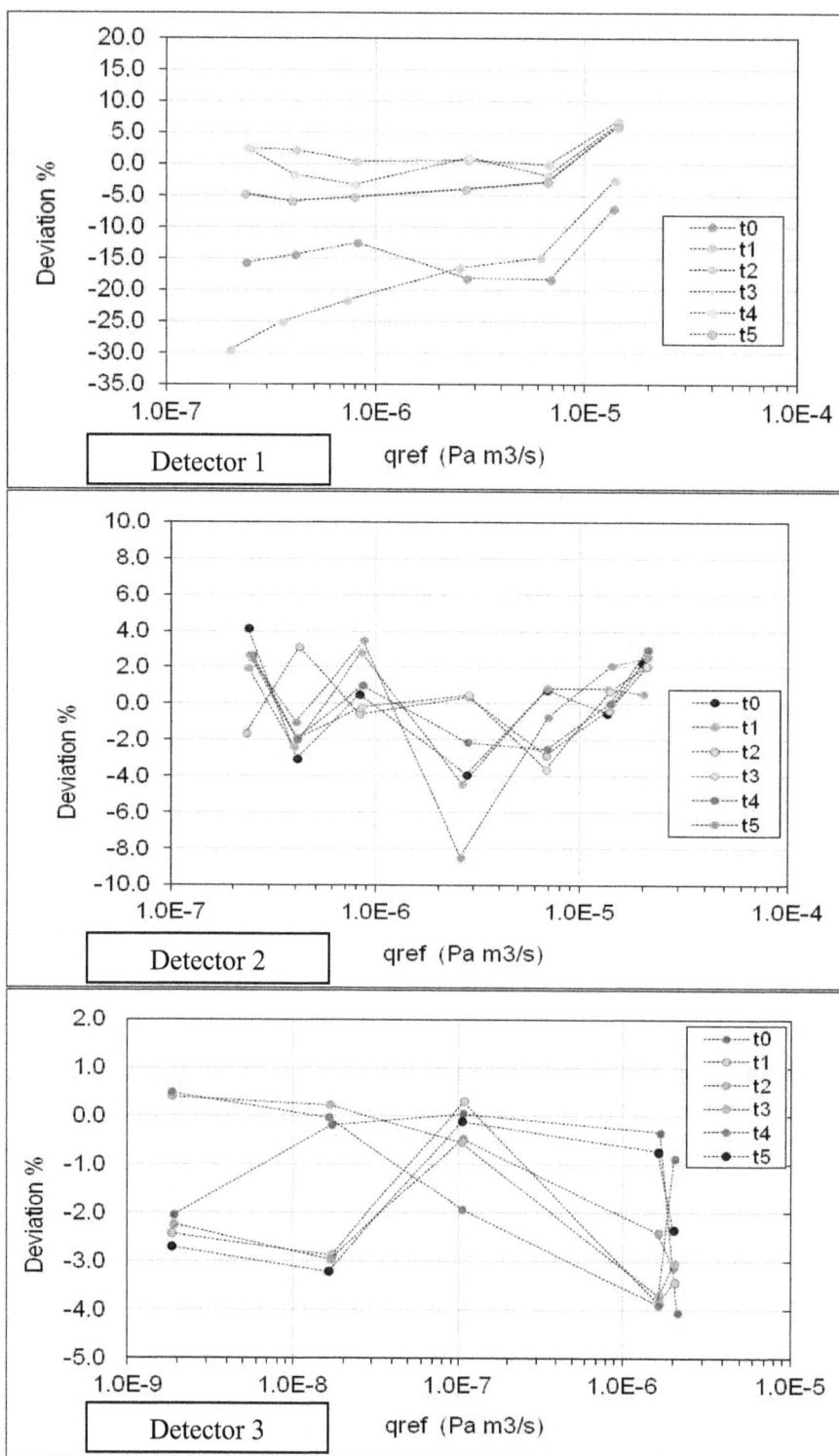

Fig 3.5. Linearity results for three helium detectors

Stability on a period of two weeks:

After the first calibration, the detectors were tested for a period of two weeks to define the stability as difference between successive readings with respect to the initial calibration at $t0$. Detector 1 (table 3.5a) has shown a stability of about $\pm$ 10% in the gas flow range considered between 10^{-7} Pa·m^3/s and 10^{-5} Pa·m^3/s.

Table 3.5a. Results of detector 1

q_{ref}	Stability at *t1*	Stability at *t2*	Stability at *t3*	Stability at *t4*
Pa·m³/s	%	%	%	%
2.4E-07	-8.2	10.1	6.7	11.1
4.1E-07	-5.7	8.5	3.9	5.7
8.1E-07	-5.6	4.7	2.1	2.0
2.7E-06	2.9	11.2	8.6	12.2
6.8E-06	2.3	10.7	9.3	9.5
1.4E-05	1.0	5.6	5.4	5.5

Detector 2 (table 3.5b) has shown, for the same gas flow directed to the detector, a stability in a range of ± 6%.

Table 3.5b. Results of detector 2

q_{ref}	Stability at *t1*	Stability at *t2*	Stability at *t3*	Stability at *t4*
Pa·m³/s	%	%	%	%
2.3E-07	-3.5	-5.7	0.3	1.4
4.1E-07	-0.5	6.5	3.0	3.9
8.3E-07	1.4	-0.3	1.4	3.5
2.8E-06	-1.0	5.8	6.8	4.8
6.8E-06	-0.1	-1.6	-1.9	-0.1
1.3E-05	1.3	2.6	3.9	3.8
2.0E-05	-1.6	2.6	2.7	4.2

Detector 3 has highlighted the best features (table 3.5c) in fact the stability is inside ±4% in a range from 2×10^{-9} Pa·m³/s to 2×10^{-6} Pa·m³/s.

Table 3.5c: Results of detector 3

q_{ref}	Stability at *t1*	Stability at *t2*	Stability at *t3*	Stability at *t4*
Pa·m³/s	%	%	%	%
1.9E-09	-0.4	-0.2	2.5	2.5
1.7E-08	-2.7	-2.8	0.4	0.2
1.1E-07	0.3	-0.5	-0.6	-2.0
1.7E-06	-3.5	-2.1	-3.4	-3.6
2.1E-06	0.6	0.9	1.0	3.2

Behavior after a self-calibration

Figure 3.6 shows the trend of the detectors one day after self-calibration at difference level of gas flow rate. All the detectors present a similar behaviour: for detector 1 and 2 the differences are inside a range between 2% and –5%; except for the point at 2.8×10^{-6} Pa·m³/s where, in case of detector 2, a difference of –8% was detected. Leak detector 3 shows a deviation inside ± 3%.

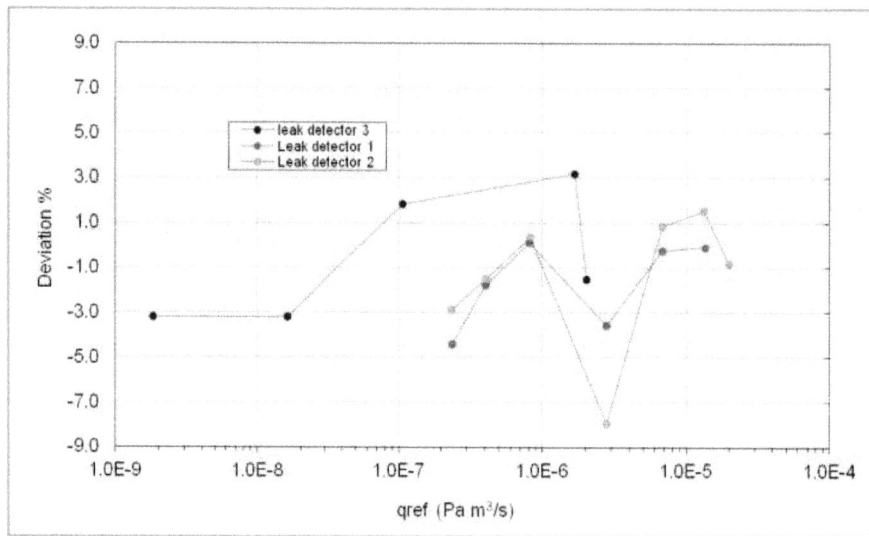

Fig.3.6. Behavior after a self-calibration of three considered detectors

III.5.4. Small leak measurement after higher leak measurement

Detector 3 was used to check the influence on the background signal due to previous higher leak measurements on the reading of a small leak. From the results of table 3.6 it can be concluded that, in this case, the effect is small.

Table 3.6. Small leak measurements after higher leak measurements

q_{ref}	Deviation for ascending measurements	Deviation for descending measurements	Difference
Pa·m³/s	%	%	%
1.9E-09	-1.8	-2.0	-0.2
1.7E-08	-2.4	-2.3	0.1
1.1E-07	0.2	-3.5	-3.7
1.6E-06	-1.6	-3.0	-1.5
2.1E-06	-2.3	-2.6	-0.3

III.6. Influence of the inlet pressure on the leak detector signal

III.6.1. Leak rates measured by secondary system in industrial laboratory, measurement referred to vacuum

Three crimped capillary reference leaks were produced and calibrated by using a secondary measurement system. The leaks were tested at two inlet pressures 99.0 kPa and 100.0 kPa to highlight the influence of the inlet pressure. The measurements were performed with the same experimental set-up and after calibration of the detector with a permeation reference leak. The signal was recorded for 30 minutes.

The measurements were repeated in two different days.

No difference in the gas flow rate was observed in the case of the biggest leak (10^{-3} mbar·L/s) changing the inlet pressure up to 1000 Pa, but the signal showed a drift during the time of measurement.

A difference of the leak detector reading of about 1.5% was found using other two capillaries for the gas flow in the 10^{-5} and 10^{-6} Pa·m^3/s range.

In table 3.6 the results are summarized.

Table 3.6. Signal stability at three different levels of gas flow rate and relative differences of the signal due to changing of the inlet pressure.

	Inlet pressure (kPa)	Leak detector reading (Pa·m^3/s)	Signal stability (%)	Relative signal difference with the pressure (%)
Leak 1	99.0	1.10E-04	1.3	0.0
Leak 1	100.0	1.10E-04	1.2	
Leak 2	99.0	1.51E-05	0.3	1.3
Leak 2	100.0	1.53E-05	0.0	
Leak 3	99.0	1.56E-06	0.0	1.3
Leak 3	100.0	1.58E-06	0.0	

III.6.2. Stability of leak rate readings at different regulated inlet pressure

The readings of the MSLD signal corresponding at four different inlet pressures of pure helium in a time of 30 minutes was recorded according to the schematic in the figure 3.7, using a secondary system referred to vacuum.

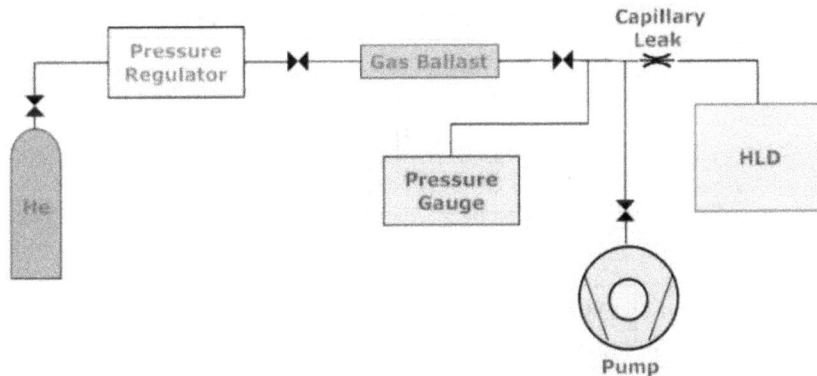

Figure 3.7. Setup to study influence of different regulated inlet pressures.

A reference capillary leak traced to primary flowmeters and a pressure regulator to maintain constant the inlet pressure were used. The measurement results are shown in table 3.7.

Table 3.7. Helium inlet pressure and its stability; gas flow rates and its stability

p_{inlet} (kPa)	p_{inlet} stability ($\times 10^{-3}$)	q (gas flow rate) Mean value (Pa·m^3/s)	Stability (%)
20	0.10	6.90E-07	0.20
40	0.05	1.39E-06	0.37
100	0.06	3.46E-06	0.43
200	0.02	7.57E-06	0.40

From the collected data, it can be pointed out that the pressure measured at the outlet of the regulator by using a digital quartz pressure sensor having good metrological characteristics (calibrated against a pressure balance $U = 3$ Pa) is stable inside about 1×10^{-4} at 20 kPa and 2×10^{-5} at 200 kPa. The helium gas flow measured in the range from 6.9×10^{-7} to 7.6×10^{-6} Pa m^3/s is also stable, in fact the signal stability, in the worst case, is equal to 4.3×10^{-3}. A similar result of the signal stability was measured by an industrial metrological laboratory as shown in the previous table 3.7.

III.7. Leak detector calibration

III.7.1. Experimental procedure to calibrate a leak detector

For quantitative measurements the leak detector must be calibrated by inletting a known helium gas flow rate into the instrument.

$$q_{He} = p_{He} S ,$$
(3.)

where q_{He} is the know helium flow rate (Pa·m^3/s), p_{He} partial pressure (pascal) and S is the pumping speed of helium (m^3/s). It is not necessary to know the pumping speed but it must be held constant during the measurements.

Before starting the calibration of the detector it is suggested to have in hand at least one reference leak per each flow rate decade with the calibration certificate and the detector manual.

An optimal set-up to calibrate the helium detector is shown in figure 3.8.

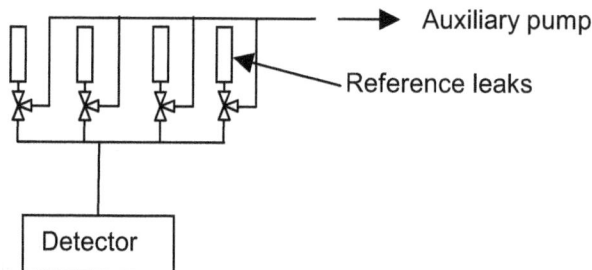

Figure 3.8. Set-up for calibration helium leak detector

The procedure can be summarized in the following steps:

1. switch on the detector, wait the sufficient time of warm up,
2. pump the leaks through a auxiliary pumping system,
3. record the background signal for 120 s,
4. connect the reference leak (starting from the smaller) to the detector directly or through a volume similar to the one that will be use during the test,
5. record the reading of the detector,
6. record the temperature of the leak,
7. repeat steps from 2 to 6 with the other leaks,
8. wait at least 1 hour,
9. repeat steps from 3 to 7 to have at least a complete second measurement cycle.

With the collected data, background noise, signal drift, linearity and the short term repeatability can be calculated.

III.7.2. Experimental procedure to calibrate a working leak

The calibration of a working leak requires a comparison between the gas flow delivered by the working leak and a gas flow realized by a reference leak having a very close value to the leak under calibration. A typical set-up to calibrate the working leak is shown in figure 3.9.

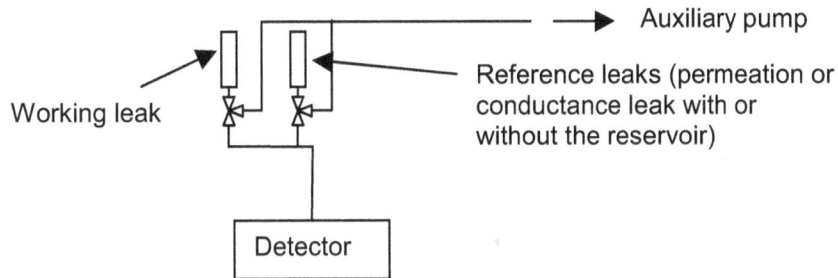

Figure 3.9. Set-up for calibration of working leak

The procedure can be summarized in the following steps:

1. switch on the detector and wait the sufficient time of warm up,
2. pump the leaks through an auxiliary pumping system,
3. record the background signal for 120 s,
4. connect the reference leak, delivering gas flow q_{ref}, to the detector,
5. record the reading of the detector (q_{read}) related to the reference leak,
6. adjust the leak detector signal of the calibration certificate following the procedure, describe in the manual to have q_{read} equal to q_{ref},
7. record the temperature of the reference leak,
8. switch off the reference leak,
9. record the background signal for 120 s,
10. connect the working leak to the detector,
11. record the reading of the detector q_{corr},
12. record the temperature of the working leak,
13. switch off the reference leak,
14. repeat steps from 3 to 11 to have at least a complete second measurement cycle.

Point 6 in the procedure is equivalent to introduce the error given by:

$$e = q_{read} - q_{ref},\tag{3.5}$$

so that, the detector has been adjusted, point 11 gives the corrected value q_{corr} of the gas flow from the working leak.

III.7.3. Uncertainty evaluation

The combined variance [21] of error can be calculating starting from equation 3.5 and is given by:

$$[u(e)]^2 = \sum_{i=1}^{2} [c_i u(x_i)]^2 = [+1 \cdot u(q_{read})]^2 + [-1 \cdot u(q_{ref})]^2 . \qquad (3.6)$$

where $u(xi)$ is a standard uncertainty of the quantity Xi and c_i are the sensitivity coefficients (see annex B). The combined standard uncertainty is given by:

$$[u(e)] = \sqrt{\begin{array}{l}[u_{resolution}(q_{read})]^2 + [u_{noise}(q_{read})]^2 + [u_{drift}(q_{read})]^2 + [u_{nonlinearity}(q_{read})]^2 + \\ [u_{repeatability}(q_{read})]^2 + [u_{driftafterselfcal}(q_{read})]^2 + [-1\ u_{ref}(q_{ref})]^2\end{array}} \qquad (3.7)$$

In the following paragraphs the various uncertainty components will be described using the results of the previous paragraphs related to detector 3.

III.7.4. Noise of the detector

The best estimate of noise is the standard deviation σ of the normal distribution of the data recorded in a period of time. The variance component due to the noise is $u^2(noise) = \sigma^2$. From table 3.1, in case of in the third measurement, u (noise) $= 1.0 \times 10^{-13}$ Pa·m³/s

III.7.5. Drift of the background signal

The drift of the background signal is defined as deviation of the signal from time zero to time t. A rectangular distribution of lower bound S_l and upper bound S_u is assumed for the deviation with associated uncertainty $u(background\ signal)=(S_u-S_l)/\sqrt{12} = (S_u-S_l) \times 0.29$. From the data of table 3.2 the drift of the signal background in the third measurement is equal to -3% which corresponds to $(S_1-S_0) = 7.7 \times 10^{-13}$ Pa·m³/s; consequently the uncertainty is:

$u(background\ signal) = (7.7 \times 10^{-13} \times 0.29)$ Pa·m³/s.

III.7.6. Resolution of the indication

One source of uncertainty is the resolution of the indicating device. If the resolution of the indicating device is δx, the value of the quantity that produces a given indication X can lie with equal probability anywhere in the interval $X-\delta x/2$ to $X+\delta x/2$, which is described by a rectangular probability distribution of width δx with variance $u^2 = (\delta x)^2/12$. The standard uncertainty of the resolution is $u = 0.29 \times \delta x$ for any indication. If the detector reading is 1.82×10^{-9} Pa·m³/s, the resolution is 0.01×10^{-9} Pa·m³/s and its uncertainty is:

$u(resolution) = 0.29 \times 1 \times 10^{-11}$ Pa·m³/s.

III.7.7. Reference gas flow

The uncertainty of reference gas flow delivered by the secondary standard leak may be deduced from the calibration certificate supplied by a National Metrology Institute.

The certificate reports the value of the gas flow delivered through the reference leak measured against a primary flowmeter having the traceability to SI units, its expanded uncertainty and the temperature of the leak during the calibration.

The expanded uncertainty, normally reported in the certificate, is the standard combined uncertainty $u_c(y)$ multiplied by a coverage factor k, typically equal to 2: $U = k\,u_c(y)$.

A summary of the expanded uncertainties of NMI's facilities is given in the table 1.2.

The standard uncertainty is evaluated by considering the uncertainty of the NMI primary standard and the short-term repeatability without taking into account the long-term stability of the reference leak.

III.7.8. Nonlinearity of the detector

The nonlinearity of the detector is defined as deviation from the linear model; all possible values of this quantity lie to one side of a single limiting value. A rectangular distribution of upper bound S_u and lower bound S_l is assumed for the deviation with associated uncertainty $u(nonlinearity) = (S_u\text{-}S_l)/\sqrt{12}$. From the data of figure 3.5c the deviation from the linearity of the detector is considered equal to -4% in the whole range; the deviation given in Pa·m^3/s for each measurement point could be calculated and the corresponding uncertainties, given by $u(nonlinearity)=(S_u\text{-}S_l)/\left(2\times\sqrt{12}\right)$, in gas flow rate units are reported in table 3.10.

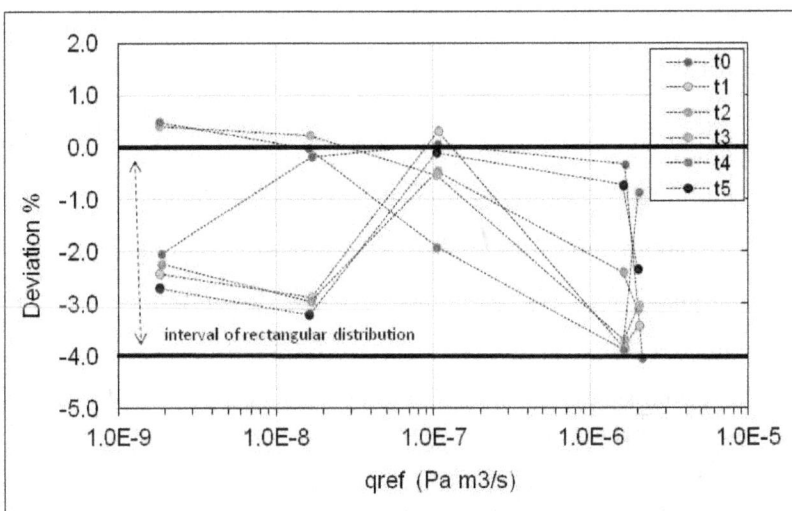

Figure 3.10. Estimate of the nonlinearity

III.7.9. Repeatability at short time 1 day:

The short term repeatability is defined as the agreement between the results of successive measurements of the same measurand carried out under the same conditions of measurement.

In many cases, the result of a measurement is determined on the basis of series of observations obtained under repeatability conditions.

Repeatability conditions include:

— the same measurement procedure

— the same measuring instrument, used under the same conditions

— the same location

— repetition over a short period of time.

The measurements of table 3.5c at time t1 are considered to explain how the short term repeatability may be evaluated. The possible values lie in the limits +0.6 % to -3.5 %. A rectangular distribution of upper bound S_u and lower bound S_l is assumed, with associated uncertainty $u(repeatability)= (S_u-S_l)/\sqrt{12}$. The (S_u-S_l) values, in gas flow rate units, and the uncertainty of measurements are given in table 3.8.

Table 3.8. Readings, intervals (4.1 %) where the possible values lie and the corresponding uncertainty

q_{read} Pa·m³/s	(S_u-S_l) Pa·m³/s	$u(repeatability)$ Pa·m³/s
1.83E-09	7.48E-11	2.2E-11
1.62E-08	6.65E-10	1.9E-10
1.04E-07	4.27E-09	1.2E-09
1.61E-06	6.60E-08	1.9E-08
2.01E-06	8.25E-08	2.4E-08

III.7.10. Drift after self calibration:

It is defined as deviation of the signal after self calibration. The possible values lie in the limits S_u and S_l; a rectangular distribution of upper bound S_u and lower bound S_l is assumed, with associated uncertainty $u(drift)= (S_u-S_l)/(\sqrt{12})$. The maximum deviation after self calibration is equal to ± 3.2% of the reading q_{read} in the whole range (figure 3.6). The (S_u-S_l) values, in gas flow rate units, and the uncertainty of measurements are given in table 3.9.

Table 3.9. Readings, intervals (6.4 %) where the possible values lie and corresponding uncertainty

q_{read} Pa·m³/s	(S_u-S_l) Pa·m³/s	u(drift after self calibration) Pa·m³/s
1.8E-09	1.16E-10	3.4E-11
1.6E-08	1.03E-09	3.0E-10
1.1E-07	6.80E-09	2.0E-09
1.7E-06	1.07E-07	3.1E-08
2.0E-06	1.30E-07	3.8E-08

III.7.11. Evaluation of combined uncertainty

Tables 3.10 and 3.11 show the uncertainty budget related to the calibration of a working leak by using detector 3 in two different gas flow levels 10^{-9} Pa m^3/s and 10^{-6} Pa m^3/s.

The expanded uncertainty of the flow from the working leak is:

$$U(q_{corr}) = 2 \cdot u(q_{corr}) = 2 \cdot \sqrt{[u(q_{read})]^2 + [u(e)]^2}, \qquad (3.8)$$

where $u(q_{read})$ is equal to the uncertainty due to the resolution which can be usually considered negligible.

Table 3.10. Example of an uncertainty budget for a working leak calibration at 1.87×10^{-9} Pa·m^3/s

q_{ref}= 1.9E-9 Pa·m^3/s $\quad$ q_{read}= 1.8E-09 Pa·m^3/s $\quad$ e = - 4.3E-11 Pa·m^3/s $\quad$ q_{corr}= 2.1E-09 Pa·m^3/s			
Uncertainty components	**Distribution of probability**	**Sensitivity coefficient c_i**	**$c_i\,u(x_i)$** Pa·m^3/s
reference flow	normal	1	4.6E-11
resolution	rectangular	-1	2.9E-12
signal noise	rectangular	-1	1.0E-13
background drift	rectangular	-1	2.2E-13
nonlinearity	rectangular	-1	2.1E-11
repeatability	rectangular	-1	2.2E-11
drift after self calibration	rectangular	-1	3.4E-11
$u^2(e)$			4.1E-21
$u(e)$			6.4E-11
$U(e)= k\,u(e)=2\,_c(e)$			1.3E-10
$U(e)/q$			7.0 %
$U(q_{corr})/q_{corr}$			7.0 %

Table 3.11. Example of an uncertainty budget for a working leak calibration at 2.04×10^{-6} Pa·m^3/s

q_{ref}= 2.0E-6 Pa·m^3/s $\quad$ q_{read}= 2.0E-06 Pa·m^3/s $\quad$ e = 0 Pa·m^3/s $\quad$ q_{corr}= 1.7E-6 Pa·m^3/s			
Uncertainty components	**Distribution of probability**	**Sensitivity coefficient c_i**	**$c_i\,u(x_i)$** Pa·m^3/s
reference flow	normal	+1	5.0E-08
resolution	rectangular	-1	2.9E-09
Signal noise	rectangular	-1	1.0E-13
Background drift	rectangular	-1	2.2E-13
nonlinearity	rectangular	-1	2.3E-08
repeatability	rectangular	-1	2.4E-08
drift after self calibration	rectangular	-1	3.8E-08
$u^2(e)$			5.0E-15
$u(e)$			7.1E-08
$U(e)= k\,u(e)=2\,_c(e)$			1.4E-07
$U(e)/q$			7.1 %
$U(q_{corr})/q_{corr}$			7.1 %

Resolution, background drift and signal noise may be disregarded in the overall uncertainty evaluation for the most common gas flow-rate range $(10^{-4} - 10^{-9})$ Pa·m^3/s; the long term stability is not included in the uncertainty budget.

III.8. Industrial application

For industrial applications [22] the leak detectors are used for approval test of various components for which the threshold value of the leaks must be defined to accept or reject the considered components. The detection process must have the shortest possible cycle time (several tens of seconds) and high reliability. The system is essentially made of a test chamber, an additional pump, an MSLD generally used in contra-flow configuration, tuned on helium, equipped with its own pumping system and reference leak, a set of valves, a gas inlet system and vacuum and pressure gauges. One additional calibrated leak (working leak) is connected directly to the object (without leak) used to adjust/test the system.

The approval threshold is *a priori* defined: the object to be tested from the production line is located inside the test chamber connected to the MSLD. After evacuation to 0.1 mbar it may be pressurized to several bar with tracer gas out of the analysis chamber or it may be hermetically closed then pressurized inside the test chamber.

The approval testing system must be characterized from the metrological point of view and the definition of threshold gas flow-rate value has to take into account the uncertainty.

If the object is leaking, the tracer gas flows from the object to the test chamber and the MSLD gives a signal. If it exceeds the threshold, the system shows an alarm, the test chamber and the object are exposed to the atmospheric pressure and the object is rejected and removed. If the signal is lower than the threshold the object is accepted and proceeds to the following production steps. At the end of each test the chamber is vented to be ready for a new cycle.

Before the calibration cycle is started, the signal of the MSLD is adjusted for the correct reading corresponding to the considered working leak by acting on the machine factor. That factor is defined as the ratio of the effective pumping speed of the pump module and the pump in the external system in measurement mode; it may be measured as the ratio between the signal corresponding to a gas flow-rate as measured when a reference leak is directly connected to MSLD and the signal obtained with the same leak connected to object or the test chamber in the test configuration.

In the test configuration the metrological characterization is the same as with the MSLD alone, but now it must be taken into account that a complete test cycle may typically last from 15 s to 40 s, time intervals during which it is necessary to evacuate and pressurize the object, to evacuate the test chamber, to detect helium and quantify it, to release the pressure from the object and vent the chamber to the atmospheric pressure. In the working configuration, the adjustment of the machine factor is considerable as it is connected to a more complex geometry and related to a shorter measurement time (3 s or 5 s) in which the steady state of the helium signal is not reached.

For the system in the working configuration, the traceability chain starts from national standards maintained by the National Metrology Institute, passes through the reference leaks to be concluded with an MSLD characterized as shown in § III.7 and maintained by the industrial metrological laboratories.

The complete traceability chain is shown in figure 3.11:

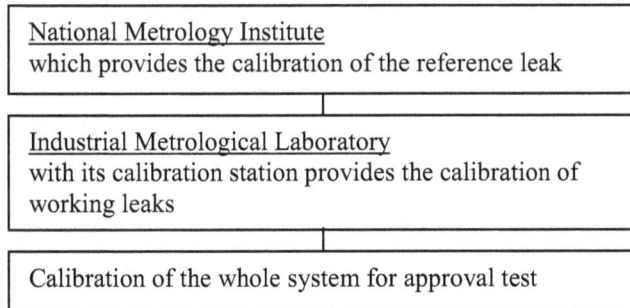

```
┌─────────────────────────────────────────────────────┐
│ National Metrology Institute                          │
│ which provides the calibration of the reference leak  │
└─────────────────────────────────────────────────────┘
┌─────────────────────────────────────────────────────┐
│ Industrial Metrological Laboratory                    │
│ with its calibration station provides the calibration of │
│ working leaks                                         │
└─────────────────────────────────────────────────────┘
┌─────────────────────────────────────────────────────┐
│ Calibration of the whole system for approval test     │
└─────────────────────────────────────────────────────┘
```

Figure 3.11. Scheme of traceability for the whole leak system

Following the previous scheme the uncertainty evaluation can be summarized

III.9. Existing standards for calibration

Non-destructive testing – Terminology - List of general terms EN 1330-1 1998
Non-destructive testing - Terminology Terms common to the non-destructive testing methods EN 1330-2 1998
Non-destructive testing - Leak testing Characterization of mass spectrometer leak detectors EN 1518 2000
Non-destructive testing - Leak testing Criteria for method and technique selection EN 1779 1999
Non-destructive testing - Leak testing Tracer gas method EN 13185 2004
Non destructive testing Leak testing Calibration of reference leaks for gases EN 13192 2002
Non destructive testing Leak testing Guide to the selection of instrumentation for the measurement of gas leakage EN 13625 2003

CHAPTER IV.   Refrigerant leak detection

Refrigerant leak detection is dedicated to air conditioning equipment in industry, automotive, *etc.*. In Europe, regulation 842/2006 gives directions, for the owners of equipment confining more than 3 kg of refrigerant fluid, to control the tightness of such equipment. In most cases, controls of leaks are performed once the equipment is filled with refrigerant, consequently one needs to process leak searching with an appropriate tool: a refrigerant leak detector (RLD). The European standard EN 14624:2012 (Performance of portable leak detectors and of room monitors for halogenated refrigerants) aims at providing series of tests to determine the detection limit (DL) which is the key characteristic of a leak detector. A DL of 5g/a according to the standard EN 14624 is a common target for refrigerant leak detectors.

This chapter presents the main physics principles of refrigerant leak detection, the tests description following the European standard EN 14624:2012 and highlights performance tests carried out on several commercially available detectors in the frame of the project JRP IND12.

IV.1. Operating principle

IV.1.1. Basic principle

The basic principle is depicted on Fig. 4.1. The gas molecules located in the volume nearby the sniffer probe are sucked through a cell (the sensor) sensitive to the refrigerant, with a rate $q_{V,d}$ (equivalent to a volumetric flow rate). The mass flow rate of refrigerant measured by the detector q_m can be deduced by the following equation [7]:

$$q_m = C \cdot \rho \cdot q_{V,d}, \tag{4.1}$$

where C is the concentration of the refrigerant measured by the sensor and ρ its density.

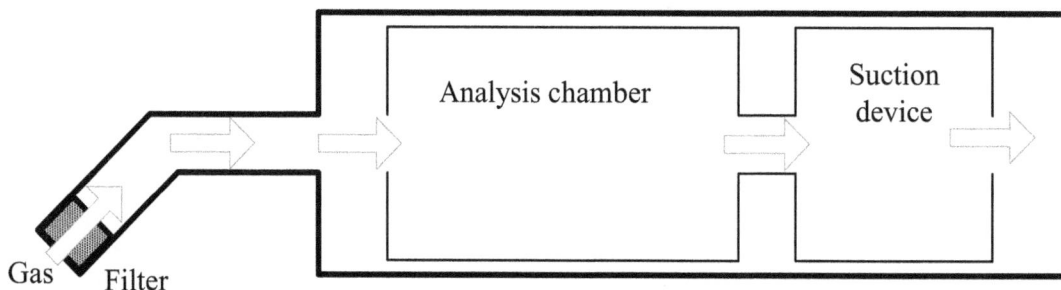

Fig. 4.1. Basic principle of a refrigerant leak detector

From equation (4.1), the minimum detectable concentration of the analyser and the suction rate of the instrument are crucial (see § I-1.4).

As concentration of the refrigerant tends to decrease with the distance from the leak's location, the mass flow rate seen by the detector is lower than the real leakage.

In most cases, refrigerant leak detectors display a light signal and/or emit a specific sound when detecting. More sophisticated instruments display a mass flow rate in g/a, either an arbitrary value or an indicated leak level by means of LEDs or bargraph.

Two main physical principles are used for sensors in commercially available RLDs: infrared absorption and electron capture (heated sensor technology).

IV.1.2. Infrared sensor

The infrared sensor is based on the absorption of infrared light by refrigerant gases. In Fig. 4.2, with refrigerant in the analysis chamber, the infrared light is absorbed so its intensity I_1, measured by the light sensor at the exit of the chamber is lower than the emitted intensity I_0. *In fine*, when the detector is switched on, the initial intensity I_0 is measured for ambient atmosphere. Furthermore, any additional presence of refrigerant gas in the analysis chamber will attenuate this intensity to I_1 and will lead the instrument to warn the user about the presence of a leak.

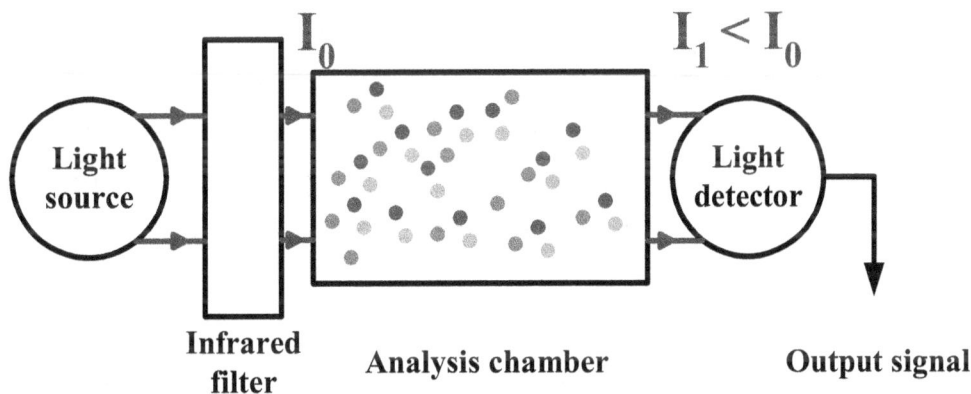

Fig. 4.2: Principle of the infrared refrigerant leak detector

IV.1.3. Heated sensor

The sensor is based on the high electro-negativity of the halogens. Therefore, detectors that use this kind of detection are composed with two elements: an emitter and a collector. The emitter is a platinum cylinder, inside which a ceramic is inserted. The collector is a platinum coil located around the emitter. When the detector is switched on, the elements are heated at more than 500 °C. The alkaline atoms migrate to the surface of the emitter and the halogens atoms capture their electrons and ionize them. Therefore a current due to the ionization of the alkaline atoms - proportional to the concentration of the halogen gas - is generated from the emitter to the collector [7].

Refrigerant leak detectors using this technology are cheaper than infrared leak detectors.

IV.2. European standard EN 14624:2012

The standard EN 14624 recommends some tests in order to assess the performance of portable leak detectors for halogenated refrigerants (and also room monitors which are not detailed in this guide). These tests, four in number, are described hereafter. The test gas is the refrigerant R-134a. The flow rate values of the calibrated leaks must be known to within ± 20 %.

IV.2.1. Technical vocabulary of the standard

The main change in the latest revision of the standard EN 14624 concerns the replacement of the term "sensitivity" by "detection limit". Indeed, "the sensitivity of a measuring system is the quotient of the change in an indication of a measuring system and the corresponding change in a value of a quantity being measured" [1] and so the word "sensitivity" was not a proper term in the context.

The refrigerant leak detection is operated in the sniffing mode (*cf.* § I.2.2): the extremity of its part used to locate leaks is the sniffer probe and the distance d between the sniffer probe and the calibrated leak (or the leaking tested object, during a detection) is the sniffing distance.

The zeroing is the operation which gives the instrument the base signal equivalent to zero leak. Consequently a leak will be detected if the measured signal varies significantly from the background signal. The zeroing is performed automatically when the instrument is switched on and can also be done by an action of the user (not for all instruments). It is important to perform the zero in the case where the atmosphere is contaminated with refrigerant; the detection limit of an instrument in such atmosphere is assessed in the test n° 3 of the standard EN 14624.

IV.2.2. Test n° 1: static detection limit

The sniffer probe is positioned stationary in front of a calibrated leak at a sniffing distance $d = (3.0 \pm 0.5)$ mm during a time that must not exceed 10 s. The text in the standard does not clearly indicate the orientation of both objects. This operation is repeated ten times. To pass the test, the detector must successfully detect the leak each time. Nominal flow rate values of the used calibrated leaks recommended by the standard are 10 g/a, 5 g/a and 3 g/a until the DL is established.

IV.2.3. Test n° 2: dynamic detection limit

The sniffer probe travels on both sides of the calibrated leak at a moving speed $(v = 2 \pm 0.2)$ cm/s, with a deviation of minimum 10 cm and a sniffing distance identical to test n° 1.

This operation is repeated ten times. To pass the test, the detector must successfully detect the leak each time. Flow rates of the used calibrated leaks recommended by the standard are 10 g/a, 5 g/a and 3 g/a until the DL is established.

IV.2.4. Test n° 3: detection limit in a contaminated environment

Test n° 3 is performed in a chamber where the concentration of R-134a can be monitored up to 1000 μmol/mol with a tolerance of ± 10 %. The detector to be tested is then switched on when the targeted concentration is reached.

IV.2.5. Test n° 4: recovery time

The sniffer probe is first placed, for 10 s, in front of the largest leak as specified by the manufacturer or a leak of 50 g/a if not specified, and then placed stationary in front of the leak corresponding to the static detection limit. The time taken by the detector to detect this leak again is measured. The test is repeated five times.

Unfortunately, in what follows, this test could not be performed because the detectors were not running properly after test n°3.

IV.3. Standard tests on commercially available detectors

The four above mentioned tests were performed in the frame of the project EMRP IND 12 [11] on eight commercially available detectors. Two of the selected detectors used the infrared technology and the other ones heated sensor technology.

IV.3.1. Description of the experimental set-up

An adjustable crimped capillary leak VTI type RLS (photo Fig. 4.3) was used to generate standard flow rates in the range between 1 and 10 g/a, with an adjustment step of 1 g/a.

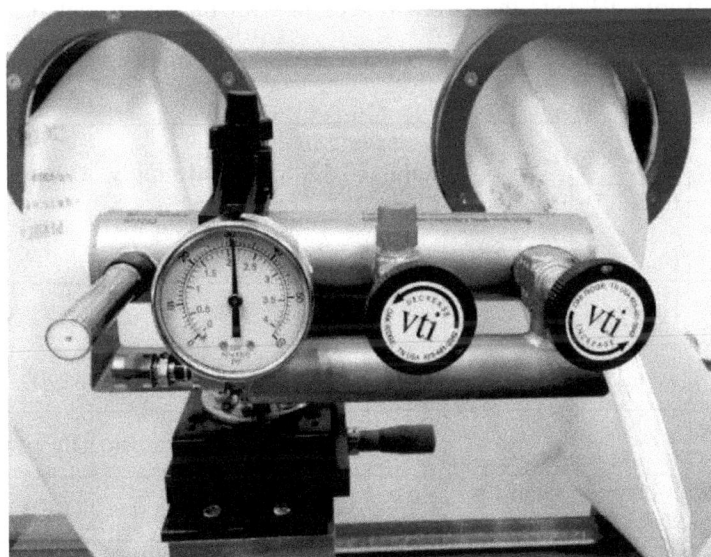

Fig. 4.3. Adjustable reference leak

The picture in Fig. 4.4 illustrates a dynamic test. The orientation of the sniffer probe is horizontal. Before the test was carried out, the sniffer probe was placed along the same axis as that of the leak artefact and the sniffing distance was adjusted at the required value $d = 3$ mm by mean of gauge blocks. The speed rate of the movable carriage can be adjusted to $v = 2$ cm/s (for tests n° 2 and 4).

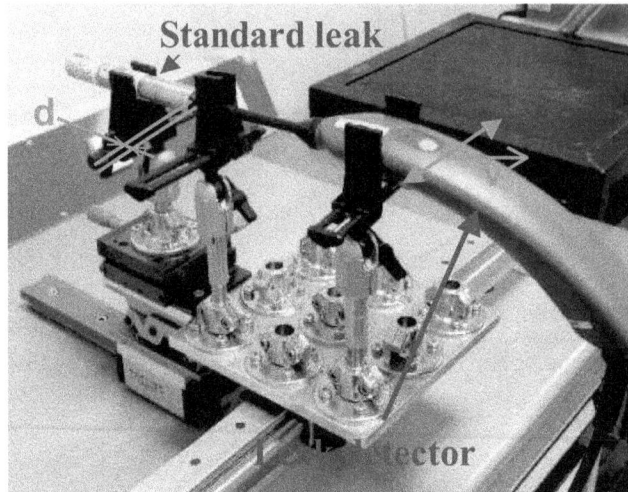

Fig. 4.4. Dynamic test performance according to the standard EN 14624:2012

d is the sniffing distance
v is the speed-rate's displacement of the detector in front of the standard leak

IV.3.2. Results

The relevant parameter is the detection limit DL. Let us suppose that a detector passes the test for a leak rate **LR₁** but fails for the next lower leak rate **LR₂** ($LR_2 < LR_1$ and $LR_2 - LR_1 = 1$ g/a). It will be stated that:

$LR_2 < DL < LR_1$.

The following Table 4.1 lists the seven detectors with their physical principle, the specified detection limit (when existing on the instruction manual), the type of electrical supplying and display.

Table 4.1. List of the tested refrigerant leak detectors

Detector #	Physics' principle	Specified DL (g/a)		According to standard EN 14624	Electrical supply	Output display
		Static	Dynamic			
1[*]	Infrared sensor	< 1	< 1	Yes	Power cord (230V or 110 V)	Digital value in g/a
2	Infrared sensor	< 1	< 1	Yes	Rechargeable battery	8 LED
3	Heated diode sensor	<2	<2	Yes	Alkaline Batteries	1 LED
4	Heated diode sensor	<1	<3	Yes	Alkaline Batteries	Digital value between 0 and 9
5	Heated diode sensor	between 3 and 14		No (SAE J1627)[**]	Alkaline Batteries	1 LED
6	Heated diode sensor	< 3		Yes	Rechargeable battery	Bargraph
7	Heated diode sensor	< 1		No [***]	Rechargeable battery	5 LED

[*] A different font color is used to distinguish detectors using infrared technology
[**] DL is given for a distance between 5 cm and the direct contact of the RLD with the leak

(***) Technical data are poorly detailed in the operating manual for this RLD

Some results are presented in the Table 4.2, 4.3, 4.4 and 4.5 regarding the tests n° 1, 2, 3 and 4 respectively (tests described in the sections 2.2 to 2.5).

Table 4.2. Detection limit determined in static conditions

Test n° 1 - Static conditions		
Detector #	Specified DL	Determined DL
	in g/a	
1	DL < 1	DL = 1
2	DL < 1	3 < DL < 4
3	DL < 2	1 < DL < 2
4	DL < 1	DL < 1
5	3 < DL < 14	6 < DL < 7
6	DL < 3	2 < DL < 3
7	1 < DL < 3	5 < DL < 6

Table 4.3. Detection limit determined in dynamic conditions

Test n° 2 - Dynamic conditions		
Detector #	Specified DL	Determined DL
	in g/a	
1	DL < 1	DL = 1
2	DL < 1	3 < DL < 4
3	DL < 2	1 < DL < 2
4	DL < 3	2 < DL < 3
5	3 < DL < 14	10 < DL
6	DL < 3	DL > 10
7	1 < DL < 3	6 < DL < 7

Table 4.4. Detection limit determined in dynamic conditions and a polluted atmosphere

Test n° 3 - Dynamic conditions in polluted atmosphere		
Detector #	Specified DL	Determined DL
	in g/a	
1	DL < 1	3 < DL < 4
2	DL < 1	4 < DL < 5
3	DL < 2	Not tested
4	DL < 3	7 < DL < 8
5	3 < DL < 14	DL > 10
6	DL < 3	DL > 10
7	1 < DL < 3	DL > 10

IV-3.3 Additional tests

The aim of the additional tests is to observe the behaviour of the refrigerant leak detectors when they are used in conditions which deviate from the test conditions of the standard EN 14624, that are likely to occur in real situations when the location of the leak is not known.

Measurements were performed with the refrigerant leak detector Inficon HLD 5000 which displays a digital value in g/a. The standard leak is the crimped capillary leak VTI type RLS (Fig. 4.3) adjusted to deliver a flow rate of 1.0 g/a.

Different orientations of the sniffer probe were tested, the distance *d* between the leak and the detector, and the speed rate *v* of the detector when it was moved in front of the leak, were varied.

The reference conditions are the ones described in the standard for the orientation, the distance *d* = 3 mm and the motion speed of the detector *v* = 2 cm/s. The results are presented via the ratio of the detector signal for the considered orientation/distance/speed motion (denoted *Signal*) to the signal measured for the reference orientation (denoted *Signal* (*ref.*)) in the static or dynamic mode. A deviation from the ratio equal to 1 is significant when it is greater than 0.05, as the display resolution of the HLD 5000 is 0.1 g/a.

IV-3.3.1 Influence of the orientation of the sniffer probe

The results obtained for the tested orientations (a-b-c-d-e) of the sniffer probe are consolidated in the following table 4.5 for both static and dynamic tests.

Table 4.5. Behaviour of a RLD for different orientations of the sniffer probe (*z-axis* is the vertical axis)

Orientation	Reference EN 14624:2012	
Static test		$$\frac{Signal(a)}{Signal(ref.)} = 0.98$$
Dynamic test		$$\frac{Signal(a)}{Signal(ref.)} = 1.01$$

Orientation	b	c
Static test	$\dfrac{Signal(a)}{Signal(ref.)} = 0.91$	$\dfrac{Signal(a)}{Signal(ref.)} = 0.94$
Dynamic test	$\dfrac{Signal(a)}{Signal(ref.)} = 0.80$	$\dfrac{Signal(a)}{Signal(ref.)} = 1.20$

Orientation	d	e
Static test	$\dfrac{Signal(a)}{Signal(ref.)} = 0.88$	$\dfrac{Signal(a)}{Signal(ref.)} = 0.92$
Dynamic test	$\dfrac{Signal(a)}{Signal(ref.)} = 0.52$	$\dfrac{Signal(a)}{Signal(ref.)} = 0.82$

IV-3.3.2 Influence of the distance in the stationary mode

The Table 4.6 shows the attenuation of the signal as a function of the distance d with respect to the signal measured for $d = 3$ mm.

Table 4.6. Behaviour of a RLD in the stationary mode when its distance from the leak is varied

Distance d	3 mm (ref.)	5 mm	7 mm
$\dfrac{Signal}{Signal(ref.)}$	1.00	0.88	0.79

IV-3.3.3 Influence of the speed rate in the dynamic mode

The Table 4.7 shows the attenuation of the signal as a function of the distance v with respect to the signal measured for $v = 2$ cm/s.

Table 4.7. Behaviour of a RLD in dynamic mode when its speed-rate related to the leak is varied; the distance is kept constant at $d = 3$ mm

Speed-rate v	2 cm/s (ref.)	3 cm/s	5 cm/s
$\dfrac{Signal}{Signal(ref.)}$	1.00	0.70	0.40

IV.3.3. Discussion on the results

Detection limit of the refrigerant leak detectors #1, 2, 3, 4 and 6 are specified according to the standard EN 14624:2012 so, in absence of any further information, it should be available for any of the three tests (n°1, 2 and 3).

All the detectors comply with the manufacturer's specification according to the standard EN 14624 in static conditions (test n°1, Table 4.3). For dynamic conditions, detectors 2, 6 and 7 do not comply (test n°2, Table 4.4). And finally, for dynamic conditions in an atmosphere polluted with refrigerant (at about 1000 ppm), none of them respond according to the standard (test n°3, Table 4.5). No statement can made as to the recovery time, since this characteristic is not given in the manuals.

A critical view of the European standard EN14624:2012 can be formulated regarding the tests performed: indeed, a single sequence is repeated ten times, and the RLD has to detect each time to pass the test successfully, which means a 100 % success. It was often observed during the tests, that an instrument failed to detect once or twice out of the ten times for the leak rate it was specified for, which is not such bad result, in reality. However, it can completely change the classification of a detector if the standard is strictly achieved, as illustrated in the following example:

Let us consider a detector **D** which is specified for 5 g/a. The laboratory which evaluates this detector, according to the standard EN 14624, has the three recommended reference leaks of 3, 5 and 10 g/a. During one of the tests, the detector **D** had 100 % success in detecting 10 g/a, but failed once in the ten sequences in detecting the leak with the rate of 5 g/a. Thus, the control laboratory states a detection limit DL: 5 g/a < DL < 10 g/a; In addition, the detector **D** does not fulfill the needs of the regulations which impose on the owners of air conditioning equipment a refrigerant detector with a detection limit lower than 5 g/a according to the standard EN14624. Notwithstanding this statement, since the detector succeeded in 90 % of the tests, it is in reality an efficient instrument for use in industry for the detection limit of 5 g/a.

IV.4. Good practices for industrial using

The ultimate good practice, available for any instrument, is to read carefully and to follow the instructions of the manufacturer.

From the principle of a detector (§ IV.1), detection is influenced by the suction speed rate of the instrument. Consequently, it will be important to:

- avoid air stream around the detector which will limit the refrigerant gas concentration through the sensor cell,

- check the battery level often, for portable detectors, as the suction requires quite a lot of energy: battery life duration in use is limited to a few hours.

When searching for leaks, users should be aware of the speed with which they move the sniffer probe. Detector tests according to the standard EN 14624 are performed at the relatively slow speed of 2 cm/s. The signal decreases rapidly when this speed is increased. It is also important to bear the short distance required between the sniffer probe and the tubing parts to be tested. However, the attenuation of the signal seems to be less critical if distance is increased compared with the speed motion (Tables 4.6 and 4.7).

The accumulation technique can be useful for small leakages. It consists in enclosing the part of equipment to be tested in a soft bag and then putting the sniffer probe in it. In this way, in case of a leakage, the refrigerant gas accumulates in the bag, so that its concentration rises. After a while, the detector is switched on. The detection is facilitated as the gas quantity was increased with the accumulation. Moreover, this technique minimises influence of draughts.

It was observed from the performed tests (Table 4.4) that detectors using the infrared technology show a better detection limit than detectors with a heated sensor, and a lower sensitivity to refrigerant contamination. In other words, such detectors provide a better level of confidence when used. Manufacturers recommend a periodicity for the replacement of the sensitive cell; in case of contamination, such an operation is necessary for heated sensor detectors.

Refrigerant contamination is a technical issue for leak detectors as it perturbs their reliable operation. When a large leak is suspected which can increase the contamination (room monitors are useful to state this), the first step of the detection is to use more adequate tools like soap-based liquids; then, when larger leaks are solved, detection with a refrigerant leak detectors can be performed.

It is difficult for users to know whether the sensor has been contaminated or not during former utilizations. So, in any case, it will be useful, even necessary for them to have a calibrated leak delivering a flow rate around 5 g/a, in order to check the instrument previously to a leak detection.

References

[1] JCGM 200:2012, International vocabulary of metrology – Basic and general concepts and associated terms (VIM), 3rd edition, 2008 with minor corrections

[2] T. Gronych, L. Peksa, P. Řepa, J Wild, J Tesař, D. Pražák, Z. Krajíček, M. Vičar, The use of diaphragm bellows to construct a constant pressure gas flowmeter for the flow rate range 10^{-7} Pa m^3 s^{-1} to 10^{-1} Pa m^3 s^{-1}, Metrologia, 45 (2008), pp. 46–52

[3] L. Peksa, T. Gronych, M. Jeřáb, M. Vičar, F. Staněk, Z. Krajíček, D. Pražák, Implementation of multi-opening orifices in the primary metrology of vacuums and small gas throughputs, Vacuum, 2013,

[4] M. Bergoglio, D. Mari, INRIM primary standard for micro gas flow measurements with reference to atmospheric pressure Measurement, 45 (2012), pp. 2459–2463

[5] M. Bergoglio, D. Mari, INRIM continuous expansion system as high vacuum primary standard for gas pressure measurements below $9 \cdot 10^{-2}$ Pa, Vacuum, 84 (1) (2009), pp. 270–273

[6] F. Boineau, Characterization of the LNE constant pressure flowmeter for the leak flow rates measurements with reference to vacuum and atmospheric pressure, PTB mitteillungen 3/2011, in: Proceedings 5th CCM International Conference, 2011, pp. 313–316.

[7] I. Morgado, Design and realization of a primary and secondary leak standards for the measurements of leak flow rates of refrigerants, Thesis, Paris, Ecole nationale supérieure des mines de Paris, 2008, 166 pp.

[8] K. Jousten, U. Becker, A primary standard for the calibration of sniffer test leak devices, Metrologia, 46 (2009), pp. 560–568

[9] K. Jousten, H. Menzer, R. Niepraschk A new fully automated gas flowmeter at the PTB for flow rates between 10^{-13} mol/s and 10^{-6} mol/s Metrologia, 39 (2002), pp. 519–529

[10] P.J. Abbott and S.A. Tison Commercial helium permeation leak standards: their properties and reliability. Journal of Vacuum Science Technology A, 13:1242-1246, (1996). [DOM1]

[11] EMRP-IND12 Vacuum: www.ptb.de/emrp/ind12-home.html

[12] H. Yoshida, K. Arai, M. Hirata, H. Akimichi New leak element using sintered stainless steel filter for in-situ calibration of ionization gauges and quadrupole mass spectrometers, Vacuum, 86, (2012), 838, 842

[13] F. Sharipov, V. Seleznev, Data on Internal Rarefied Gas Flows J. Phys. Chem. Ref. Data, Vol. 27, No. 3, 1998

[14] NIST webbook http://webbook.nist.gov/chemistry/fluid/

[15] M. A. Gallis, J. R. Torczynski, Direct simulation Monte Carlo-based expressions for the gas mass flow rate and pressure profile in a microscale tube, Physics of fluid 24, 012005, (2012)

[16] A. Nerken, History of helium leak detection, J. Vac. Sci. Technol. A, Vol. 9, No.3, May/Jun 1991

[17] P. E. Mix, Leak Testing Methods, Introduction to Non destructive Testing: A Training Guide, Second Edition, John Wiley & Sons, Inc., Hoboken, NJ, USA.

[18] Handbook Non-destructive testing, Third edition, published by American Society for non-destructive Testing, 1998

[19] Introduction to Helium Mass Spectrometer Leak Detection, ed. Varian Vacuum Products, 1995.

[20] K. Jousten, Handbook of vacuum technology, Wiley-VCH, 2008

[21] Evaluation of measurement data – Guide to the expression of uncertainty in measurement JCGM 100:2008 http://www.bipm.org/en/publications/guides/#gum

[22] A. Calcatelli, M. Bergoglio and D. Mari, Leak detection, calibrations and reference flows: practical example, Vacuum 81,11-12 (2007) 1538-1544

[23] U. Becker, D. Bentouati, M. Bergoglio, F. Boineau, W. Jitschin, K. Jousten, D. Mari, D. Prazak, and M. Vicar, 'Realization, characterization and measurements of standard leak artefacts', *Measurement*, vol. 61, pp. 249–256, 2015.

[24] http://iopscience.iop.org/0026-1394/50/1A/07001K Jousten, et al; Final report of key comparison CCM.P-K12 for very low helium flow rates (leak rates); Metrologia, 50, Technical Supplement 2013

Annex A- Table of fluid isothermal data for helium

Temperature (C)	Pressure (bar)	Density (kg/m3)	Volume (m3/kg)	Internal Energy (kJ/mol)	Enthalpy (kJ/mol)	Entropy (J/mol*K)	Cv (J/mol*K)	Cp (J/mol*K)	Sound Spd. (m/s)	Joule-Thomson (K/bar)	Viscosity (Pa*s)	Therm. Cond. (W/m*K)	Phase
20.0	0.1000	0.016421	60.897	3.6769	6.1143	130.84	12.472	20.786	1007.5	-0.062299	1.9614e-05	0.15343	vapor
20.0	0.6000	0.098504	10.152	3.6770	6.1150	115.95	12.472	20.786	1007.7	-0.062300	1.9616e-05	0.15347	vapor
20.0	1.1000	0.18055	5.5387	3.6770	6.1156	110.91	12.472	20.785	1007.9	-0.062300	1.9618e-05	0.15351	vapor
20.0	1.6000	0.26256	3.8087	3.6771	6.1163	107.79	12.473	20.785	1008.1	-0.062301	1.9620e-05	0.15355	vapor
20.0	2.1000	0.34452	2.9026	3.6772	6.1169	105.53	12.473	20.785	1008.3	-0.062301	1.9622e-05	0.15359	vapor
20.0	2.6000	0.42645	2.3449	3.6773	6.1176	103.76	12.473	20.785	1008.6	-0.062302	1.9624e-05	0.15363	vapor
20.0	2.6000	0.42645	2.3449	3.6773	6.1176	103.76	12.473	20.785	1008.6	-0.062302	1.9624e-05	0.15363	supercritical
20.0	2.6000	0.42645	2.3449	3.6773	6.1176	103.76	12.473	20.785	1008.6	-0.062302	1.9624e-05	0.15366	supercritical
20.0	3.1000	0.50834	1.9672	3.6773	6.1182	102.29	12.474	20.785	1008.8	-0.062303	1.9625e-05	0.15370	supercritical
20.0	3.6000	0.59020	1.6944	3.6774	6.1189	101.05	12.474	20.785	1009.0	-0.062303	1.9627e-05	0.15374	supercritical
20.0	4.1000	0.67201	1.4881	3.6775	6.1195	99.969	12.475	20.785	1009.2	-0.062304	1.9629e-05	0.15378	supercritical
20.0	4.6000	0.75379	1.3266	3.6776	6.1202	99.013	12.475	20.784	1009.4	-0.062304	1.9631e-05	0.15382	supercritical
20.0	5.1000	0.83552	1.1969	3.6776	6.1208	98.155	12.475	20.784	1009.6	-0.062305	1.9633e-05	0.15385	supercritical
20.0	5.6000	0.91722	1.0902	3.6777	6.1215	97.378	12.476	20.784	1009.9	-0.062305	1.9635e-05	0.15389	supercritical
20.0	6.1000	0.99888	1.0011	3.6778	6.1221	96.667	12.476	20.784	1010.1	-0.062306	1.9637e-05	0.15393	supercritical
20.0	6.6000	1.0805	0.92549	3.6779	6.1228	96.012	12.476	20.784	1010.3	-0.062306	1.9638e-05	0.15396	supercritical
20.0	7.1000	1.1621	0.86052	3.6779	6.1234	95.405	12.477	20.784	1010.5	-0.062307	1.9640e-05	0.15400	supercritical
20.0	7.6000	1.2436	0.80409	3.6780	6.1240	94.840	12.477	20.784	1010.7	-0.062307	1.9642e-05	0.15404	supercritical
20.0	8.1000	1.3251	0.75463	3.6781	6.1247	94.310	12.477	20.784	1010.9	-0.062307	1.9644e-05	0.15408	supercritical
20.0	8.6000	1.4066	0.71093	3.6782	6.1253	93.813	12.478	20.783	1011.2	-0.062308	1.9646e-05	0.15411	supercritical
20.0	9.1000	1.4881	0.67202	3.6782	6.1260	93.343	12.478	20.783	1011.4	-0.062308	1.9648e-05	0.15415	supercritical
20.0	9.6000	1.5694	0.63717	3.6783	6.1266	92.899	12.479	20.783	1011.6	-0.062308	1.9650e-05	0.15419	supercritical
20.0	10.100	1.6508	0.60577	3.6784	6.1273	92.477	12.479	20.783	1011.8	-0.062309	1.9651e-05	0.15419	supercritical

Annex B- Uncertainty: main definitions

The following definitions are from the internation vocabulary of metrology [1].

quantity: property of a phenomenon, body, or substance, where the property has a magnitude that can be expressed as a number and a reference

derived quantity: quantity, in a system of quantities, defined in terms of the base quantities of that system. Example: in a system of quantities having the base quantities length and mass, pressure is a derived quantity defined as the quotient of mass, gravity and area.

dimension of a quantity: expression of the dependence of a quantity on the base quantities of a system of quantities as a product of powers of factors corresponding to the base quantities, omitting any numerical factor

unit of measurement: real scalar quantity, defined and adopted by convention, with which any other quantity of the same kind can be compared to express the ratio of the two quantities as a number

measurement unit that is adopted by convention for a base quantity

measurement: process of experimentally obtaining one or more **quantity values** that can reasonably be attributed to a quantity

measurement result, a **measurement procedure**, and a calibrated **measuring system** operating according to the specified measurement procedure, including the measurement conditions.

measurand: quantity intended to be measured.

error: measured quantity value minus a reference quantity value.

systematic measurement error: component of measurement error that in replicate measurements remains constant or varies in a predictable manner.

random measurement error: component of measurement error that in replicate measurements varies in an unpredictable manner

repeatability condition: condition of measurement, out of a set of conditions that includes the same measurement procedure, same operators, same measuring system, same operating conditions and same location, and replicate measurements on the same or similar objects over a short period of time
measurement repeatability: measurement precision under a set of repeatability conditions of measurement

reproducibility condition of measurement condition of measurement, out of a set of conditions that includes different locations, operators, measuring systems, and replicate measurements on the same or similar objects

measurement reproducibility measurement precision under reproducibility conditions of measurement

measurement uncertainty: non-negative parameter characterizing the dispersion of the quantity values being attributed to a measurand, based on the information used

Type A evaluation of measurement uncertainty evaluation of a component of measurement uncertainty by a statistical analysis of measured quantity values obtained under defined measurement conditions from the statistical distribution of the quantity

Type B evaluation of measurement uncertainty: evaluation of a component of measurement uncertainty determined by means other than a Type A

measurement uncertainty: expressed as a standard deviation The **uncertainty of measurement** is a non negative parameter, associated with the result of a measurement that characterises the dispersion of the values that could reasonably be attributed to the measurand

combined standard measurement uncertainty: standard measurement uncertainty that is obtained using the individual standard measurement uncertainties associated with the input quantities in a measurement model

relative standard measurement uncertainty: standard measurement uncertainty divided by the absolute value of the measured quantity value.

uncertainty budget: statement of a measurement uncertainty, of the components of that measurement uncertainty, and of their calculation and combination

expanded measurement uncertainty: product of a combined standard measurement uncertainty and a factor larger than the number one

coverage interval: interval containing the set of true quantity values of a measurand with a stated probability, based on the information available.

coverage factor: number larger than one by which a combined standard measurement uncertainty is multiplied to obtain an expanded measurement uncertainty

calibration: operation that, under specified conditions, in a first step establishes a relation between the quantity values with measurement uncertainties provided by measurement standards and corresponding indications with associated measurement uncertainties and, in a second step, uses this information to establish a relation for obtaining a measurement result from an indication

measurement model: mathematical relation among all quantities known to be involved in a measurement

measurement function: function of quantities, the value of which, when calculated using known quantity values for the input

influence quantity: quantity that, in a direct measurement, does not affect the quantity that is actually measured, but affects the relation between the indication and the measurement result

Annex C- Uncertainty assessment general consideration

The result of a measurement after correction for recognized systematic effects is still only an estimate of the value of the measurand because of the uncertainty arising from random effects and from imperfect correction for systematic effects [21].

In most cases, a measurand Y is not measured directly, but is determined from N other quantities X_1, X_2, ..., X_N through a functional relationship f:

$$Y = f(X_1, X_2, ..., X_N).$$

(C.1)

The standard uncertainty of y, where y is the estimate of the measurand Y, is obtained by combining the standard uncertainties of the input estimates $x_1, x_2, ..., x_N$.

The combined standard uncertainty $u_c(y)$ is the positive square root of the combined variance $u_c^2(y)$, which is given by:

$$u_c^2(y) = \sum_{i=1}^{N} \left(\frac{\partial f}{\partial x_i} \right)^2 u^2(x_i) = \sum_{i=1}^{N} [c_i u(x_i)]^2 ,$$

(C.2)

where f is the function given in Equation (C.1). Each $u(x_i)$ is a standard uncertainty and the combined standard uncertainty $u_c(y)$ characterizes the dispersion of the values that could reasonably be attributed to the measurand Y.

The partial derivatives $\partial f/\partial x_i$, called sensitivity coefficients, describe how the output estimate y varies with changes in the values of the input estimates $x_1, x_2, ..., x_N$.

Uncertainty of measurement includes, in general, many components. Some of these components may be evaluated from the statistical distribution of the results of series of measurements and can be characterized by experimental standard deviations. The other components are evaluated from assumed probability distributions based on experience or other information.

The uncertainty of measurement associated with the input estimates is evaluated according to 'Type A' or a 'Type B' method of evaluation. The Type A evaluation of standard uncertainty is the method of evaluating the uncertainty by the statistical analysis of a series of observations. In this case the standard uncertainty is the experimental standard deviation of the mean that follows from an averaging procedure. The Type B evaluation of standard uncertainty is the method of evaluating the uncertainty by means other than the statistical analysis of a series of observations. In this case the evaluation of the standard uncertainty is based on some other scientific knowledge.

Type A evaluation of standard uncertainty

The Type A evaluation of standard uncertainty can be applied when independent observations have been made on the some input quantities under the same conditions of measurement. With n statistically independent observations ($n > 1$), the estimate of the quantity Q is the arithmetic mean of the individual observed values q_j ($j = 1, 2, ..., n$), $\overline{q}$.

The uncertainty of measurement associated with the estimate is given by:

$$s^2(q) = \frac{1}{n-1} \sum (q_j - \bar{q})^2 \,.$$ (C.3)

The standard uncertainty $u(q)$ is its positive square root:

$$u(q) = s(q) \,.$$ (C.4)

Type B evaluation of standard uncertainty

The Type B evaluation of standard uncertainty is the method of the uncertainty associated with an estimate x_i of an input quantity X_i by means other than the statistical analysis of a series of observations. The standard uncertainty $u(x_i)$ is evaluated by scientific judgment based on all available information on the possible variability of X_i. When only a single value is known for the quantity X_i, e.g. a single measured value, a resultant value of a previous measurement, a reference value from the literature, or a correction value, this value will be used for x_i. The standard uncertainty $u(x_i)$ associated with x_i is to be adopted where it is given.

When a probability distribution can be assumed for the quantity X_i, based on theory or experience, then the appropriate expectation or expected value and the square root of the variance of this distribution have to be taken as the estimate x_i and the associated standard uncertainty $u(x_i)$, respectively.

If only upper and lower limits $a+$ and $a-$ can be estimated for the value of the quantity X_i (e.g. manufacturer's specifications of a measuring instrument, a temperature range), a probability distribution with constant probability density between these limits (rectangular probability distribution) has to be assumed for the possible variability of the input quantity X_i.

$$x_i = \frac{1}{2}(a_+ + a_-),$$ (C.5)

for the estimated value and:

$$u^2(x_i) = \frac{1}{12}(a_+ + a_-)^2 \,.$$ (C.6)

If the difference between the limiting values is denoted by $2a$:

$$x_i = \frac{1}{3}a^2 \,.$$ (C.7)

The rectangular distribution is a reasonable description in probability terms of one's inadequate knowledge about the input quantity X_i in the absence of any other information than its limits of variability.

The expanded uncertainty U is obtained by multiplying the combined standard uncertainty u_c by a coverage factor k, $U = k\, u_c$. The expanded uncertainty U provides an interval about the result of a measurement that may be expected. To obtain the value of the coverage factor k that produces an interval corresponding to a specified level of confidence requires detailed knowledge of the probability distribution characterized by the measurement result and its combined standard uncertainty. The factor k is usually equal to 2 or 3 based on the coverage probability or level of confidence required. $k{=}2$ defines an interval estimated having a level of confidence of 95.45 percent and $k{=}3$ an interval having a level of confidence 99.73%.

The uncertainty of measurement characterizes the dispersion of the values that could reasonably be attributed to the measurand. The uncertainty analysis for a measurement, should include a list of all sources of uncertainty together with the associated standard uncertainties of measurement and the methods of evaluating them. For repeated measurements the number n of observations also has to be stated. For clarity, data relevant to this analysis are presented in the form of a table. In this table all quantities should be referenced by a physical symbol X_i or a short identifier. For each of them at least the estimate x_i, the associated standard uncertainty of measurement $u(x_i)$, the sensitivity coefficient c_i and the different uncertainty contributions $u_i(y)$ should be specified. The measurement unit of each of the quantities should also be stated with the numerical values given in the table.

A formal example of such an arrangement is given in the following Table C.1 applicable for the case of uncorrelated input quantities. The standard uncertainty associated with the measurement result $u(y)$ given in the bottom right corner of the table is the root sum square of all the uncertainty contributions in the outer right column.

Table C.1. Schematic of uncertainty contributions used in the uncertainty analysis of a measurement.

Quantity X_i	Estimate x_i	Standard uncertainty $u(x_i)$	Probability distribution	Sensitivity coefficient c_i	Contribution to the standard uncertainty $u_i(y)$
X_1	X_1	$u(x_1)$		c_1	$u_1(y)$
X_2	X_2	$u(x_2)$		c_2	$u_2(y)$
X_N	x_N	$u(x_N)$		c_N	$u_N(y)$
Y	y				$u(y)$

Annex D- Simple model to predict the gas flow from a leak

The development of the theory of gas flow in narrow ducts is an objective related to improvement of leak testing. At present, the leaks generally used in the industrial environment are obtained by crimping small stainless steel capillaries in order to generate gas flow in the 10^{-5} mbar L/s – 10^{-3} mbar L/s range suitable for leak testing [10]. Due to the unknown geometry and flow characteristic an empirical description of gas flow rate is difficult to achieve. Therefore, leak elements with well-defined geometry were realized in the framework of the project EMRP-IND12.

Three different materials were used to realize the leak artefacts: stainless steel, copper and aluminium, with the aim of defining a convenient, inexpensive material in which regular holes of definite thickness can be produced using laser ablation.

The artefacts have been characterized: the diameter of the hole and its thickness were carefully measured by using systems traced to INRIM standards. The flow rate delivered by the leak was measured with primary standards in the National Institutes participating to the project. The measurements were carried out against vacuum and atmospheric pressure for different gases to cover the various gas regimes inside the ducts.

From the data collection a simple model to describe the regime inside the ducts from molecular to viscous was applied. This could help the leak testing operator to predict the gas flow in other conditions (in particular, different gases), starting from few measurements performed with only one species of gas.

The available experimental data have been compared with the predictions of the bridging formula proposed by Gallis and Torczynski [15]. The general expression is:

$$\dot{M} = \dot{M}c\left(1 + \frac{8p_\lambda}{p_m}\overline{\omega}[p_1, p_2]\right). \tag{D.1}$$

The quantities appearing in the previous equations are defined as follows: $\dot{M}$ is the mass flow rate, $\dot{M}c$ is the mass flow that would be achieved at continuum conditions, p_m is the average of the inlet p_1 and outlet p_2 pressure, and $\overline{\omega}$ is a dimensionless function:

$$\overline{\omega}(p_1, p_2) = \frac{2-\alpha}{\alpha}\left\{(1 + b_1\alpha) + (\varepsilon b_0 - (1 + b_1\alpha))\frac{b_2 p_\lambda}{p_1 - p_2}\ln\left[\frac{p_1 + b_2 p_\lambda}{p_2 + b_2 p_\lambda}\right]\right\}, \tag{D.2}$$

with:

$$p_\lambda = \frac{p\lambda}{D} = \frac{\pi\mu c}{4D}, \tag{D.3}$$

$$p_m = \frac{p_1 + p_2}{2}, \tag{D.4}$$

$$\dot{M}c = \left(\frac{D^4 p_m (p_1 - p_2)}{16\mu c^2 L}\right). \tag{D.5}$$

c is the molecular thermal speed, λ is the molecular free path and μ the dynamic viscosity. The terms εb_0, b_1 and b_2 determine the behaviour of the mass flow rate in the free- molecular, slip and transition regime:

$$b_0 = \frac{16}{3\pi},$$
(D.6)

$$b_1 = 0.15,$$
(D.7)

$$b_2 = 0.7\frac{\alpha}{2-\alpha},$$
(D.8)

$$\varepsilon = \frac{1+K}{\delta + K},$$
(D.9)

$$K = \frac{\delta - 1}{\delta}\frac{\alpha L}{D},$$
(D.10)

$$\delta = (2-\alpha)\frac{4}{3}.$$
(D.11)

p_1 and p_2 represent the pressure at the entrance and at the exit of the channel, they are not in general equal to the pressures in the tanks $p_{1\,\infty}$ and $p_{2\,\infty}$, consequently they must be determined. Following the discussion of the paper [15] p_1 and p_2 can be determined in the first approximation in the following way:

$$p_i = q_i - 6p_\lambda, \ (i=1,2),$$
(D.12)

where

$$q_i = \sqrt{\frac{(1+F)(p_{i\infty} + 6p_\lambda)^2 + F(p_{j\infty} + 6p_\lambda)^2}{1+2F}}, \ (i,j) = (1,2)\,;(2,1)$$
(D.13)

and

$$F = \frac{3\pi D}{32L}\left[1 + \frac{16p_\lambda}{p_{1\infty} + p_{2\infty} + 12p_\lambda}\left(\overline{\omega}(p_{1\infty}, p_{2\infty}) - \frac{3}{4}\right)\right].$$
(D.14)

The model described was applied to leaks realized in the framework of the project EMRP- IND12 and the results were compared with the experimental measurements carried out in different National Metrology Institutes [23]. For the leaks considered and for each gaseous species two sets of experimental data are available: in the first set, the molar flow was measured by setting outlet pressure very close to zero (expansion into vacuum); the measurements of the second set were obtained by setting the outlet pressure close to 10^5 Pa (expansion at atmospheric pressure).

From the first set of experimental points a wide range of inlet Knudsen numbers from free molecular to viscous regime was highlighted. Hence, the experimental data have been compared with the values evaluated by the previous equations that can cover the whole range of rarefaction conditions.

Flow conditions in the second set of data fall in the final part of the transition regime. For such flow conditions bridging formulas would give the same results as applying the Hagen-Poiseuille approximation.

The inlet and outlet hole diameters were measured, but the inner channel geometry is known only approximately. An effective circular channel diameter *Deff* has been determined from each experimental data set by minimizing the deviation between calculated and experimental data.

Figure D1 summarizes both experimental data [23] and calculated values of the normalized conductance ($C_n = C \cdot \sqrt{m}$) versus the inverse of the mean free path while in Figure 2.11 the relative differences between experimental and calculated data are plotted.

The maximum relative difference of 20 %, between the experimental data and the values coming from the bridging formula have been found Figure 2.12. It corresponds to the highest values of the inlet pressure of the leak or to a situation in which a polyatomic gas as R-134a has been expanded at atmospheric pressure.

A simple EXCEL program to predict the gas flow delivered by a leak with the model described previously is available at http://martin.vicar.cz/ind12/

Annex E- *International community acceptance of calibration certificate*

In international trade there is an increasing demand for demonstration of traceability to the SI system of all the measurement processes. Such traceability, generally, is proved by the calibration certificates issued by a NMI or by accredited laboratories. The validity of a certificate is guaranteed by the reliability of the laboratory that has issued the certificate and by the actual visibility of the traceability chain.

That happens briefly, by two steps:

- SI traceability

- Key comparison

The traceability is automatically guaranteed directly by using a primary device (e.g. primary flowmeter), otherwise, transfer leaks characterized with reference to a primary flowmeters existing in the NMIs.

Even if a complete chain is guaranteed as regards SI traceability and the related uncertainties are evaluated, only comparisons between metrology laboratories give enough reliance about the actual comparability of the gas flow rate generated by different systems within the stated uncertainty values. One of the main aims of intercomparisons is to indicate a way of identifying possible systematic effects not already taken into account in the uncertainty budget.

To establish the equivalence of the national standards for the quantity of our interest, the Comité International des Poids et Mesures (CIPM), through the Comité Consultatif *pour la Masse et les Grandeurs Apparentées* (CCM), has promoted several comparisons of fundamental importance (key comparisons); the Regional Metrological Organizations (RMO), among them EUROMET, have also began several intercomparison projects in which laboratories equipped with transfer gauges traceable to other NMI can also participate.

In October 1999, CIPM promoted the mutual recognition arrangement (MRA) signed by 38 NMIs, whose objectives are "to establish the degree of equivalence of national measurement standards maintained by NMIs; to provide for the mutual recognition of calibration and measurement certificates issued by NMIs". Such objectives may be achieved through the international comparisons that are the above mentioned key comparisons and supplementary comparisons.

A very important part of the MRA concerns the quality system adopted by a NMI which must be clearly indicated and meet the requirements of ISO 17025 standard.

Among the various requirements of the ISO 17025 standard there is a complete clear demonstration of the SI traceability of all the measurements, comprising those involved in the primary evaluation of the flow rate values. This means, among other requirements, that for all the quantities, not only the base ones (e.g. mass, length, time…), calibration certificates issued internally or by an external body must be made available. This complete traceability makes not only calibrations and measurements performed by a NMI reliable but also those performed by a laboratory accredited by it. Complete demonstrable traceability accompanied by appropriate procedures can provide an easy way of checking that all the necessary steps have been carefully fulfilled.

To enable national metrology institutes providing service for leak rate calibrations to apply for these entries in the data base and to ensure international equivalence in this field, the key comparison CCM.P-K12 was organised [24]. The goal of this comparison was to compare the national calibration standards and procedures for helium leak rates. PTB´s vacuum section acted as pilot laboratory. Two helium permeation leak elements of 4×10^{-11} mol/s (L1) and 8×10^{-14} mol/s (L2) served as transfer standards and were measured by 11 national metrology institutes for L1 and 6 national metrology institutes for L2. Since the calibration measurement capability of each laboratory was to be tested, it was decided that the necessary temperature environment for the leak artefact was to be provided by each participant.

The measurand determined by each laboratory was the molar flow rate q_n of helium molecules flowing out of the transfer standard leaks at the time of calibration.

In the following figures E1 and E2, the results of the pilot laboratory and all participants are shown for both the leak measured in the comparison.

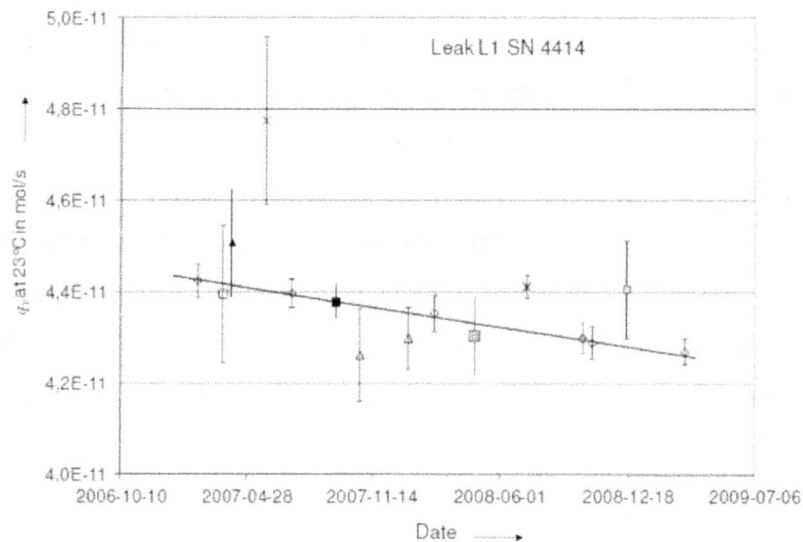

Figure E.1. Results of world-wide comparison between 11 national laboratories (the pilot laboratory carried out several measurements to check the stability of the leak L1)

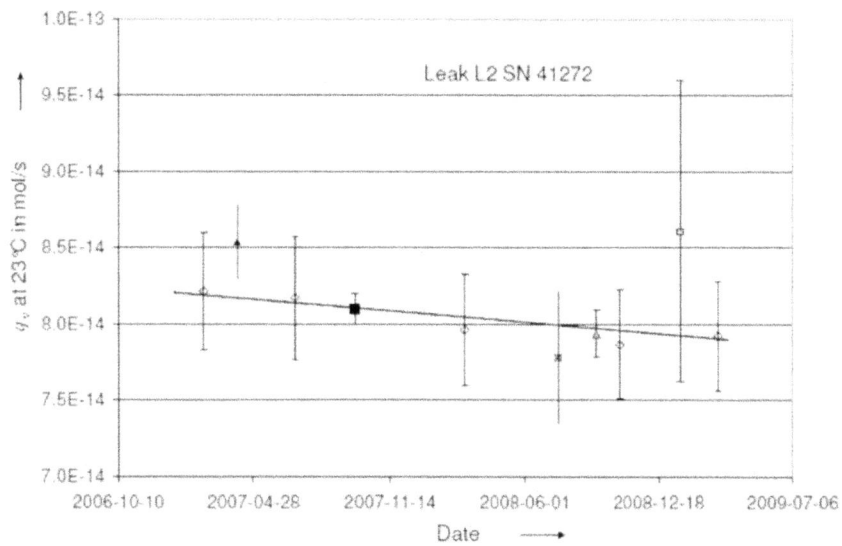

Figure E.2. Results of world-wide comparison between 6 national laboratories (the pilot laboratory carried out several measurements to check the stability of the leak L2)

Equivalence was shown for 8 laboratories in the case of L1 and for all 6 in the case of L2.

www.ingramcontent.com/pod-product-compliance
Lightning Source LLC
Chambersburg PA
CBHW082307210326
41598CB00028B/4466